Conflicts in Construction

T0315357

Conflicts in Construction

Avoiding, Managing, Resolving

Jeffery Whitfield
Executive Director
Hill International (UK)

A John Wiley & Sons, Ltd., Publication

This edition first published 2012
© 2012 John Wiley & Sons, Ltd.
First edition published in 1994 by Macmillan Press Ltd

Wiley-Blackwell is an imprint of John Wiley & Sons, formed by the merger of Wiley's global
Scientific, Technical and Medical business with Blackwell Publishing.

Registered Office
John Wiley & Sons, Ltd, The Atrium, Southern Gate, Chichester, West Sussex, PO19 8SQ, UK

Editorial Offices
9600 Garsington Road, Oxford, OX4 2DQ, UK
The Atrium, Southern Gate, Chichester, West Sussex, PO19 8SQ, UK
2121 State Avenue, Ames, Iowa 50014-8300, USA

For details of our global editorial offices, for customer services and for information about
how to apply for permission to reuse the copyright material in this book please see our
website at www.wiley.com/wiley-blackwell.

The right of the author to be identified as the author of this work has been asserted in
accordance with the UK Copyright, Designs and Patents Act 1988.

Library of Congress Cataloging-in-Publication Data

Whitfield, Jeffery, author.
  Conflicts in construction / Jeffery Whitfield, Director, Hill International (UK).
    pages cm
  Includes index.
  ISBN 978-1-118-29870-1 (pbk.)
1. Construction industry–Management. 2. Conflict management. I. Title.
  HD9715.A2W46 2012
  624.068′4–dc23
                                            2012015857

A catalogue record for this book is available from the British Library.

ISBN: 978-1-118-29870-1 (pbk.)

Wiley also publishes its books in a variety of electronic formats. Some content that appears
in print may not be available in electronic books.

Cover design by Sandra Heath
Cover photo courtesy of Shutterstock

Set in 10/12.5pt Minion by SPi Publisher Services, Pondicherry, India

1 2012

Contents

Preface: A Time of Change

I have been writing this book in the run up to 2012, an exciting and a challenging time for the country and the construction industry. Whilst there is a strong demand, and need, for new construction of all types, there is also a strong yet unfulfilled demand for the funding of that construction growth.

In the UK, the last government did not meet its promises on affordable and decent housing. New school and college buildings, along with large capital projects in the health sector, were promised but unfunded. Under the austerity measures introduced by the new government, even infrastructure spending has suffered, and all of this is happening as governments around the world are looking to reduce their spending too.

We can read in the printed media and watch on TV as it becomes clear that the UK is not alone in reducing government spending. I travel widely to assist in international construction disputes and have seen tower cranes standing idle across China, Southeast Asia and in India, the very countries that are said to be booming. The Middle East is continuing to spend on infrastructure and industrial projects, but my recent visits have coincided with major commercial and residential project suspensions, deferments and cancellations. There is considerable doubt about whether many of the United Arab Emirates' ambitious plans for new 'cities' will come to fruition soon, or ever. Their plans for spectacular theme parks, projected to dwarf even Disney World in Florida, are slowing dramatically, and lenders are questioning the wisdom of placing enormous city-sized theme parks in both Dubai and in Abu Dhabi, when they are only an hour apart by road.

The good news is that some sectors are holding up very well. New retail capital spending, mainly large supermarket chains, is on the rise. Power companies are building on a scale not seen for many years, to try to make up for a lack of investment over the past three decades, in an effort to reduce carbon emissions. Oil, gas and chemical investment has slowed a little but they too are looking to the future, 2015 to 2020, when demand will again outstrip supply.

The challenges are great, the opportunities greater, but what will these straitened circumstances mean for conflict in the construction related industries? Read on and together you and I will try to forecast what shape conflicts will take, but I suspect that many of the historic causes of conflict will continue to repeat themselves for the foreseeable future.

1 Conflicts in Construction

The term 'Built Environment' describes the product of what is probably the most diverse industry in the world, 'the Construction Industry'. Not only does the built environment cover a wide variety of end products from apartment blocks to oil refineries, but the people working within the construction and engineering industries are drawn from a broad range of trades and professions. With architecture and almost every branch of finance, engineering and surveying represented, there can be 20 or more professional disciplines involved in a moderately complex construction project.

The projects tackled by the industry are not only wide ranging and varied in their end use, but they also have significant variety within their types, for example; a power station can be fuelled by oil, gas, coal, bio fuel or nuclear products and each type of power station is very different in design, size, output and cost. The proposition that construction produces more variety than any other industry is supported by the premise that almost every major building project is unique, a prototype, a one-off.

So it is that for every complex project that stands to be constructed, a learning curve is inevitable. It is a rare industry indeed that produces so many varied prototypes without significant repetition.

The construction industry is further complicated by the presence of numerous parties with an interest in the completed structure. These include the end user, the funding parties, the developer, the planning authority, the construction regulators, the Health and Safety Executive and the public at large, whose built environment is important to them economically, environmentally and aesthetically. Add to these divergent interests the expectations of the Contractor who builds the asset, his sub-contractors and suppliers who contribute specialist plant, skills and materials, and there are relatively few people in our society without at least a small investment in the construction process.

It has been argued that it is the high number of interested parties within the construction process that provides the catalyst for conflict in the industry, and we know that disputes in construction are common. Construction conflicts

Conflicts in Construction: Avoiding, Managing, Resolving, Second Edition. Jeffery Whitfield.
© 2012 John Wiley & Sons, Ltd. Published 2012 by John Wiley & Sons, Ltd.

range from the very private two-party differences over the meaning of a contract clause, to the very public outcry over proposals for a development to be sited in countryside defined as being of outstanding natural beauty. Such conflicts can be addressed and managed only when we understand the true causes of the conflict. Whilst an understanding of how the industry is organised is necessary, on its own it is not enough.

From this point forward, any reference to the construction industry will encompass the civil engineering, heavy engineering, marine and offshore sectors too, as in my experience in these sectors all conflicts tend to sprout from the same roots.

1.1   Recent history

There have been many changes in the Built Asset industry over the last 40 years. Perhaps the most dramatic of these changes has been the sharp increase in the incidence of serious conflict between the parties to construction contracts. In 1960, some 250 writs were issued relating to construction disputes, yet within 30 years, this number increased five-fold. What happened within those intervening years to bring about such a transformation?

Perhaps it was due to the fact that during this period companies became more commercially aware. Increased numbers of quantity surveyors were employed by constructors to safeguard their contractual rights and the rights of sub-contractors. Since the 1980s, even sub-suppliers have recognised the need for sound commercial advice. Furthermore, new contract conditions have proliferated, ensuring that even experienced practitioners have become uncertain of the terms embodied in the standard form contracts.

In the late 1960s and early 1970s, we saw the emergence of claims consultants, who were established solely to promote the claims advanced by both sides of the industry. In the 1980s, there was a move away from significant public financing of construction by central and local government, and so the industry became reliant upon profit-oriented, speculative development. The early 1990s brought us the Private Finance Initiative (PFI) and other private funding routes for major projects. Since 1997, public expenditure has increased again and many more publicly funded assets were being constructed, or promised, until the General Election of 2010.

One of the most worrying aspects of the built environment for investors, practitioners and end users has been the tendency of the construction economy to react disproportionately in periods of boom and bust. Thus the cyclical nature of financial and political investment has disrupted the gradual and stable growth of the construction industry, which was a hallmark of post-war industrial and commercial development.

All of these factors have undoubtedly impacted on the quality of the relationships enjoyed between end users and their contractors (and sub-contractors). General informed opinion is that these uncertainties have also increased the incidence of conflict.

In addition to the problems within the construction industry itself, society as a whole has encountered other wide-ranging social changes that inevitably influence our attitudes to conflict. These cannot be ignored and they are addressed later in this book.

With conflicts wreaking havoc on the time and cost outcomes of major projects, onshore and offshore, we can no longer sit back and tinker with our contracts and our commercial arrangements, hoping that the resulting decline in conflict will be as rapid as was the growth. Intensive action is needed now to manage the conflicts that seem to arise inevitably from virtually every construction project.

1.2   Understanding conflict

Before we can take the radical measures that are needed to manage conflict successfully, we must be able to define the nature, identity and causes of conflict. The first step in this process will be to recognise that the difficulties the construction industry has faced over the past 40 years are not the real causes of construction conflict, but are merely symptoms. To examine the real causes fully, we need to address four major issues:

1.   Can we avoid unnecessary conflict?
2.   How do we control inevitable conflict?
3.   How do we manage conflict, whilst awaiting a resolution?
4.   How do we finally resolve construction conflicts?

We can all enhance our skills by seriously studying these vital issues. Our ability to answer these four questions will benefit not only ourselves, but also the whole industry, leading ultimately to substantial savings in unnecessary dispute costs and to better relationships, as more conflicts are avoided or are resolved amicably.

1.3   Addressing conflict

The key to successful conflict management is a fuller understanding of the various aspects of conflict, including how and why conflict arises. We must acknowledge that conflict is more than simple disagreement and we need to accept that throughout our lives we can expect to encounter conflict in one form or another.

For the purposes of classification, a strict definition of conflict will require us to decide whether we are defining *conflict* the noun or *conflict* the verb. It is my view that *a conflict* (noun) can itself be defined in three different ways.

1.   Firstly, it is a state of opposition or hostilities, for example a war or a battle. It might also be described as a fight or more subjectively defined as a struggle for freedom.

2. A second definition considers a conflict to be a clashing of opposed principles; this may be more conveniently described as a conflict of beliefs.
3. The third definition is the opposition of incompatible wishes or needs within an individual and the stress or distress resulting from this. Psychologists describe this in their terms as internal conflict; or we might call it a conflict of values.

Unfortunately, in the construction industry, there has been a tendency to ignore these three definitions and to restrict our definition of *conflict* to its use as a verb, thus overlooking real opportunities to positively resolve conflicts in the most appropriate way. The verb *to conflict* is defined as:

- to struggle; or contend;
- to contradict.

Whilst the definitions of *conflict* as a verb or a noun appear to have only a negative aspect, positive results can emanate from conflict, both in the outside world and in a construction setting.

1.4   Positive aspects of a conflict

Our own individual personalities, fundamental as they are to our existence as human beings, will usually involve us in a struggle to achieve what we see as being worthwhile. The freedom to choose allows us to pursue what we consider to be worthy causes.

Most of the aspirational management books of the twentieth century sought to improve our behaviour as human beings by offering us success or by encouraging us to pursue loftier ideals or goals, goals which would benefit society as well as ourselves. The titles of these books were provocative, thoughtful and often challenging. Some of my favourites were:

- *Think and Grow Rich*; Napoleon Hill;
- *How to win friends and influence people*; Dale Carnegie;
- *The One Minute Manager*; Ken Blanchard;
- *The 7 Habits of Highly Effective People*; Stephen R Covey;
- *The Dilbert Principle*; Scott Adams.

Well, perhaps not so much the last one, although it contains at least as many truths as the others. So, what do we learn when we struggle to achieve our lofty aspirations? Generally, we learn something more about ourselves; we become stronger, we progress.

Even the most destructive conflicts can bring forward the greatest acts of courage and heroism, displaying the very best of human potential. Winston Churchill recognised this in his renowned war time speech where he said:

Never in the field of human conflict was so much owed by so many to so few.

Recently in Egypt, conflict has also been the prime tool of the revolutionary in overthrowing oppressive governments. Positive conflict has enabled civil rights movements to remove inequitable laws. Conflict can also produce positive aesthetic results. Many of our great works of art have been created as a result of the internal conflicts raging within talented individuals. One such example could be Vincent Van Gogh. Creative individuals who challenged established thinking were often regarded as eccentric or even borderline insane. Why is it then that conflict, often so damaging, can sometimes produce such positive results?

There is within the human persona a natural tendency to fight for what we believe to be just or to defend that which we regard as being of value. This desire to seek justice and to defend our beliefs can develop those talents that were previously hidden. Conflict can also rebuild long lost community spirit. Communities can be brought together in a unique way when the people concerned have a common purpose. This purpose becomes the mortar that holds together a group of otherwise disparate individuals. Examples of this include local protection groups fighting to prevent – or bring about – a new development, such as a new road or bypass. Consider the long-running conflict fought by the people opposing the new development at Heathrow. In such a case, even if the cause is lost or the motion defeated, the benefits of working towards the common purpose can be seen to result in a closer-knit community.

Conflict also has a tendency to build teamwork and interdependency. When facing sublime opposition, we group together for comfort and for strength, we bond and create relationships that last. I have always been moved by the US Marines motto:

No man left behind

In recent conflicts in Afghanistan and Iraq, our own soldiers have followed this principle to bring badly injured or dead soldiers out of harm's way. When we fight together for a common purpose, we cannot do it alone; we must become truly interdependent and thus, dependent. When we rely on someone or need someone, we instinctively become closer to them and even protective towards them. On the other hand, we can come to resent those upon whom we depend. Understanding these variable but instinctive human behaviours can help us better comprehend and thus avoid the causes of conflict.

If we were to make it our goal to actively seek to create a lasting relationship by pursuing a common purpose, rather than looking for areas of disagreement, we might transform our future negotiations at a stroke. Furthermore, if we entered into discussions with the objective of reaching an outcome that helped the other side as much as it benefited us, we might make better progress.

So, is it essential that conflict generally, and construction conflict in particular, is eliminated? Is it always the best policy to eliminate conflict?

Clearly, in many instances, conflict can and should be eliminated. In the construction industry, a great many destructive and costly conflicts have arisen that could have been avoided, a recent example being the construction of the new Wembley Stadium completed in 2007, disputes still running in 2010.

If all deadlines were met, if all invoices were paid in full and on time, if all specifications were realised, if all claims were reasonable and honest, if all expectations were realistic, and if the people involved communicated with accuracy and complete understanding, with a willingness to be flexible and to seek solutions to problems that would benefit all concerned, conflict within the industry would be rare. Of course, with so many variables, and human nature being a factor, a degree of conflict is inevitable.

However, there are ways in which we can eliminate unnecessary conflict, and keep that conflict which is unavoidable, to the minimum. Our ability to manage a conflict successfully will depend largely upon our recognition of the real causes. Many of the perceived causes of conflict, as we have previously noted, are only symptomatic of a more fundamental underlying problem. To treat the symptoms of construction conflict is to give some temporary relief, whilst overlooking the need for a permanent cure.

1.5   The real causes of conflict

The true causes of conflict within the construction industry are wide ranging and varied, and these will need to be discussed in more detail later. However, we can highlight some of the more common problems here in summary form.

1.5.1   *Misunderstandings*

Conflicts often arise through misunderstanding. These misunderstandings usually involve some element of poor communication. For example, we probably all know that in order to achieve something worthwhile we must have a goal, a timescale and some standards or rules to which we must conform. Unfortunately, it is often the case that managers send their subordinates off to carry out a task without really explaining what they expect them to achieve; either that or the subordinates set off without having listened attentively to their instructions.

I have a son who, when he was younger, enjoyed helping me by doing simple tasks. On many occasions I would say, 'Son, would you please just go upstairs and bring…' By the time I reached the latter part of the sentence he was gone and could be heard searching around upstairs – for what, he had no way of knowing. Within a few minutes he would come back more slowly than he went and would ask sheepishly 'What did you want from upstairs?'

Communication is a two-way mechanism. It needs a listener and a speaker. It also requires the listener and speaker to change places occasionally, becoming speaker and listener. Many attempted communications do not manage to communicate at all. So often, instead of the dialogue that is necessary, we are confronted with two parties speaking monologues alternately, or worse, in unison. If you want a living example of this principle, listen to BBC Radio's *Today* Programme, when two opposing politicians are 'conversing'. A wise man once said, 'If you want to see eye to eye, try using your ears.'

As a young quantity surveyor, working on a new school building, I encountered this principle personally when I was faced with what could have become a serious problem. The floor tiling supplier ceased trading before the finishes were started, the specialist non-slip tiles specified by the Architect were only available from one other source and the delivery time for new orders was a minimum of 12 weeks. The new school term was only six weeks away. Delay was unthinkable. The Architect panicked and instructed the Contractor to quickly find, procure and install some alternative non-slip tiling for the entrance area. If possible the colour was to be black. A week later I found the Architect and Contractor locked in a bitter quarrel. The new tiling contractor had supplied grey quarry tiles with a non-slip surface. They met the non-slip requirement and the time requirement and, the Contractor explained, they were not available in black, but as this had seemed to be only a preference, he did not think it critical. Luckily, threats of claim and counter-claim quickly subsided when, during an inspection visit, the school's head teacher expressed her delight at the new building and in particular the welcoming grey tiled foyer.

Once individuals are made cognisant of what they are expected to achieve in quantitative and qualitative terms, it is for their managers or supervisors to specify the rules and standards that are to apply in the execution of the task. A failure to specify rules and standards can also cause a conflict, because we may be left with an end product that meets our needs but which has left trouble in its wake. There is an anecdote that explains the principle better than I ever could:

> The scene is England in the early Middle Ages and the King is trying to unify his country under one banner. As he sits awaiting news of progress, one of his trusted messengers enters his chamber. 'Sire. We have secured the West Country. Our men cut a swathe through your enemies, pillaged their lands and took their young women.' The king looked puzzled before replying, 'I have no enemies in the West Country.' The messenger paused and then answered with solemnity, 'You have now.'

Having set goals, and clearly communicated the rules and standards to be observed, we have a duty to explain the resulting consequences of failure. Both legally and morally we should specify the outcome that can be expected if the goal is not achieved. The attempted imposition of some consequence for failure, not previously disclosed, will certainly cause unnecessary conflict and will, quite reasonably, engender a sense of injustice.

So, to avoid conflict through misunderstanding, we must communicate with precision. How this can best be achieved is discussed in detail in a later chapter.

1.5.2   *Sensitivity*

As with all of the readers of this book, I am a rational, sensible and well balanced individual, except for those occasions when I am none of these things. Later in the book, I discuss the events that can impact on our sense of balance, our rationality and our reasonableness.

There are days when an individual's mood will change with a rapidity that may scare partners, friends and co-workers. We are probably all aware from life experience that these mood swings can have an effect on an individual's demeanour, patience and tone of voice. It may also impact on their judgement when they are supposed to be considering the needs of others.

There are many truly wonderful individuals who are passive and genteel on most occasions, but whose life challenges have left them permanently sensitive to specific topics of conversation.

I was once in a heated site meeting when a colleague, concerned that the sub-contractor's representative was becoming angry, vocally expressed his view that there were no reasons for tempers to flare. The exact expression he used was: 'Keep your rug on, no-one has any intention of holding up your payments.' I am quite certain that the MD of the sub-contracting organisation did not hear the second part of that sentence. A rather obvious wearer of a toupee, the MD's hand involuntarily flew to his head at the words 'Keep your rug on,' moving his hairpiece left and right until it was back where it should have been. There are tragic moments in life where you know that laughter is not the best medicine, and yet even the most controlled individuals find it difficult to suppress a giggle of embarrassment. The sub-contractor's team walked out and my colleague threw his head into his hands and assured us that his colloquialism was nothing more than an unintended subliminal remark for which he would apologise promptly. Sensitivity can be found in almost every environment, even on testosterone fuelled building sites.

As a young QS, I worked with a 'Ganger', an expression probably not used today, but this labour supervisor was the toughest man I ever met. He often expressed the opinion that he regarded office employment as being reserved for the weak of mind and body. One of his great pastimes was to humiliate the new 17-year-old site QS (me) in front of as many operatives as possible, a pastime he undertook with relish and considerable success. After a few days away from work, I was sitting alone in the site office, which in those days was little more than a glorified garden shed, when he came in and began to tease me about my absence. Before he could embarrass himself further, I explained that my absence had been due to the fact that my Mother had just died, and I waited to see whether my obvious pain would temper his teasing. I was not expecting his reaction. I hoped that he would remain silent before expressing his condolences. He was, after all, a genuinely nice guy under the bravado, but what did happen shocked me. This strapping hulk of a man sat down on a chair and burst into tears. He sobbed uncontrollably for five minutes. His back story explained all. He had adored his widowed Mother and treated her like a queen. He had lived at home well into his thirties. Then one day he received a message to call head office from the site telephone in an office much like mine, only to hear of his beloved Mother's sudden demise. My story and sadness had resonated with him.

We should not become paranoid about upsetting others whose sensitivities may not be obvious, but we should be aware that conflict can arise if we happen to touch a raw nerve.

By pure coincidence, as I was writing this section, I had the need to make a significant purchase by first testing the market with a handful of suppliers. I have read and re-read my enquiry email many times in the light of what happened next and I still cannot see where my text was ill advised. Having stated that the product must be the same quality as the sample cited and must include the same support, I advised the suppliers of the limited budget that was available, asking if they could supply in accordance with my request. I would not have been offended had they declined to make an offer of supply. The first response I received was along these lines:

> Of course we can meet that quality, we have been in this business for years and we are somewhat surprised that you feel the need to ask if we support our products properly. We do. Anyway, we cannot meet your price, we are not here to subsidise your profits. If you wish to increase the budget let us know.

In line with usual policy I wrote to all parties explaining that one of them had indeed provided the goods as specified indicating the price range. My guess is that the errant supplier had experienced a frustrating day at the office and that by the time my email arrived he was ready for a fight.

Sensitivity runs in both directions and so we should always be careful to consider whether we too are being over sensitive in inferring insults where none are expressed or intended. A safe way to avoid over reacting in such cases is to maintain our values at all times, even when provoked.

1.5.3 Values

Many of the contributors to a successful construction or engineering project are professional men and women. These individuals will not only have high personal values, but they will also be expected to exercise a high standard of professional ethics.

Architects, quantity surveyors, engineers and PMs have codes of practice, which denote minimum standards of behaviour. Failure to meet these standards may result in discipline or expulsion from their professional body.

These values can, when a crisis occurs, cause internal conflict. An architect friend of mine worked consistently for a major developer. His style of architecture suited the developer and the developer remained loyal to my friend, despite strong fee competition from other practices. At the time of this story, he was engaged in work on a new retail park. The project suffered from a poor pre-letting rate and the developer, himself under pressure from his banks, asked his architect to undervalue the Contractor's work, find reasons to withhold the completion certificate or do whatever was necessary to keep the developer solvent. The Architect faced an ethical crisis. He undoubtedly owed the developer a great deal, yet his personal and professional values required him to be fair and impartial in certifying the value of work done. Which route would he choose? On this occasion, he wisely chose to ignore the needs of his client and certify honestly. He confided to me later, 'I lost a good client that day, but in fairness I would not have wanted to go on working for him if he expected me to betray my own values.'

Ethical choices, often with significant sums of money attached, are a part of the construction professional's life, and learning to manage internal conflict can improve our ability to make wise decisions. The methodology leading to that improvement will be explained later.

1.5.4 Interests

We can sometimes confuse our real interest in an outcome with an unrealistically high expectation, and thus introduce a conflict where none should exist. Whilst we all appreciate that we cannot have our cake and eat it, as human beings we do try to hedge our bets as far as possible.

A client may want a quick, quality building at a low price and a contractor may want to take more time, reduce quality and get as much as he can for it. In such extreme cases their interests diverge, but they need not do so. Both might well be satisfied with a quality building built reasonably quickly at a fair price. Unfortunately, we sometimes adopt the wrong solution to a problem, with the result that a conflict of interests arises unnecessarily.

There are ways of avoiding and reducing this type of conflict, and we will look at these methods in due course.

1.5.5 People

The personalities of the individuals involved in a project will often cause them to react to one another in a destructive way.

Some people become emotional and defensive when they lack information. Other individuals will make poor judgements under stress, rather than wait for a time when they can think more rationally. Managers often try to demonstrate dominance over their staff, when the real problem is a lack of self-esteem within themselves. Those involved in a conflict often adopt a hard adversarial stance, because they do not want to be perceived as weak for adopting a conciliatory approach.

The anthropology and psyche of the human race affects the type and quantity of conflict in the construction industry and so a fuller understanding of both is necessary to identify the real causes underlying the surface behaviour of the individual. I will examine these aspects of human behaviour in a later chapter.

In this overview, I have highlighted some of the more common problems that lead us into conflict, but much more needs to be said on each topic before we can avoid or manage conflict to create a more harmonious industry.

2 Why Do We Need to Manage Conflict?

When conflict has been adequately defined and the reasons for its occurrence within the industry are discerned, we can move forward and look at ways of successfully managing that conflict. The way we manage a conflict will usually be different for each incident that we encounter. Each type of conflict has a unique cause and so, in order to reach a satisfactory outcome, each type of conflict will have a unique solution.

Persuading people, who operate in what is perceived as a somewhat macho industry, that conflict needs to be managed at all, can be difficult. Many experienced individuals genuinely believe that conflict brings benefits to the industry. Some PMs even consider themselves to be the beneficiaries of conflict. These men – and they mostly are men – have a powerful presence and a strong personality that enables them to override the legitimate concerns and rights of others. In practice, this strong approach serves only to push the conflict underground and the battle will eventually be fought at another time and another place. The clarion call of these individuals appears to be:

Conflict is absolutely necessary. Without it we just couldn't get the job done.

Others in the industry take a more pragmatic view of conflict and genuinely see no real hope of avoiding a conflict. To these people the phrase:

Construction is naturally contentious and you won't change that

trips easily off the tongue. Whilst this is clearly recognisable as a defeatist attitude, it is still surprisingly difficult to overcome.

These two views, that conflict is useful and that it is unavoidable, can lead to a relaxed approach to controlling conflict. So, how are we to convince these individuals, who often hold influential positions in the industry, that managing conflict is both necessary and beneficial? Firstly, we must explain the types of conflict, and secondly, we must explain the avoidable results of an uncontrolled conflict.

Conflicts in Construction: Avoiding, Managing, Resolving, Second Edition. Jeffery Whitfield.
© 2012 John Wiley & Sons, Ltd. Published 2012 by John Wiley & Sons, Ltd.

Management theorists have concluded, quite reasonably, that there are two types of construction conflict, namely functional conflict and dysfunctional conflict.

2.1 Functional conflict

This is a term applied to a conflict that results in progress and achievement or in a better outcome than would otherwise have been expected. It may also be described as positive or productive conflict. To a point, there is some value in this theory, but a functional conflict must still be properly managed. By allowing a functional conflict to run its natural course, we take the risk of it maturing into a dysfunctional or counterproductive conflict.

2.2 Dysfunctional conflict

This term is applied to a type of conflict that prevents progress, avers achievement and suspends success. It is a destructive conflict. No real benefits are claimed for this type of conflict and so the theorists argue that controlling or managing this type of conflict is essential. Unfortunately, very few people are able to properly categorise functional and dysfunctional conflicts at the time the conflict arises. The categorisation of conflict into functional and dysfunctional is almost inevitably done retrospectively by the uninvolved.

So, in practical terms, we have no real choice. We must manage every conflict, whether it turns out to be functional or dysfunctional, from the outset. We simply cannot foresee the eventual outcome of a conflict whilst it is in its embryonic stage.

History shows us that seemingly insignificant conflicts can rapidly escalate into major contentions. Who would have guessed that the assassination of a relatively obscure Archduke by the name of Ferdinand, on a stone bridge in Bosnia, would have been the catalyst for the Great War in 1914?

History is littered with minor events that diverted the course of human progress – sometimes positively, sometimes negatively. The problem seems to be that no-one really knows whether a conflict is going to be a force for good or ill in advance of the event. As noted above, we can of course avoid having to anticipate an uncertain future by controlling and managing every conflict and managing it properly from its inception. The alternative is to cross our fingers and hope that everyone involved in a functional conflict will continue to behave rationally, reasonably and remain unemotional – a prospect we would be ill advised to gamble upon.

We have all experienced minor disagreements that have turned into major arguments in the home, at work and even on the playing fields of our favourite sport. In these cases, the real issues are often forgotten in the heat of battle and the dispute deteriorates into an emotive personality clash, often resurrecting old unresolved problems. When this happens, it is quite natural for one party to wish to hurt the other side in some way. In this damaging, and often escalating, mode of conflict, nobody really wins. There are usually two battered and bruised losers.

The management and control of every conflict is absolutely necessary to avoid this type of unhappy situation prevailing. As a body, we need to move away from the lose/lose situations of the past and also away from the win/lose scenarios that have been so common in the industry. We can then identify and pursue the win/win outcome that would be so beneficial for construction. Some cynics will argue that there can never be two winners and that one side must always lose or must at least give something up in order to obtain an agreement or a settlement. Whilst generally this may appear to be true, it is not necessarily so. Inventive and forward looking negotiators have often increased the size of the cake, so that both parties were able to get a bigger slice. This type of invention and innovation cannot easily be achieved in a 'them and us' environment. It requires both parties to work together towards a mutually satisfactory outcome. Whilst this is by no means the traditional way of resolving a construction conflict, it may well prove to be the preferred route once the alternatives have been considered. The prospect that a conflict may turn out to be dysfunctional or damaging to the project should be enough to convince the cynics that managing conflict is essential. However, if more reasons are needed, they do exist and they are equally potent.

In a conflict, people often become psychologically stressed to such a degree that many seek medical help to overcome anxiety or to help them weather traumatic periods at work. Anxiety is described, by the medical profession, as the feeling we encounter when we are frightened. The physical signs of anxiety include:

- racing heart rate or palpitations;
- dizziness or light-headedness;
- butterflies in the stomach;
- trembling hands;
- dry mouth;
- flushes and sweating;
- wanting to go to the toilet;
- rapid breathing;
- difficulty swallowing;
- the urge to fight or flee.

Anthropologists tell us that these are natural responses to danger and that they emanate from our early antecedents who really needed them to avoid attacks by ravening beasts. Of themselves, these effects do not pose a threat to our physical health, but we are also advised that the more frequently we subject ourselves to such anxiety, the more susceptible we become in the future. In extreme cases, we can suffer irrational and seemingly unrelated panic attacks long after the conflict is resolved.

In researching this section, I consulted a doctor in general practice within a suburb of a large town, largely populated by professionals, to ask about the effects of stress on managers. During the course of the interview, it became clear that GPs are now offering 'Beta Blockers', a form of tranquilliser, to businessmen who would otherwise be overly anxious about attending meetings or negotiations. If our conflicts are regularly causing us to suffer to the extent where we need medical intervention, is this not a good enough reason for managing conflicts or avoiding them altogether? If we do not control conflict, then we may be faced with those symptoms of stress that are potentially dangerous – high blood pressure, anxiety, sleeplessness, inability to relax or settle, snappiness or bad temper, loss of interest in pleasurable activities and hobbies (including sex) and the development of nervous habits such as tics. Any one of these can make life miserable, but some of them are effectively life threatening, especially those that carry physiological symptoms such as high blood pressure.

We should not overlook the psychological pressures either, as these have led many businessmen to find that the enjoyment they derived from living was no longer sufficient to maintain their interest in going on and subsequently they have felt it necessary to take their own lives. One of the saddest days of my business life came when a client company, a family run business, rang to tell me that one of the directors had taken his own life because of a raging conflict with one of his main customers. This type of stress must be reduced and it can only be avoided by the proper control of conflict.

One extremely persuasive reason for the early management of conflicts is to save the time and effort that unresolved conflicts will otherwise consume. Conflicts take weeks, months or even years to run their course naturally, and during this time relationships become strained or they simply disintegrate. This occurs most often because the real issues are put to one side, unresolved, while the parties continue to deal with the urgent day-to-day problems of running the contract. Many arbitration agreements choose to defer the resolution of conflict until the end of the project, and this can result in disharmony, lack of commitment and poor performance during the currency of the job.

Another problem with leaving the resolution of live conflicts until the end of the job is that further damage may accrue during the remainder of the project, and so the final bill may be considerably increased. Furthermore, at the end of the job, the claimant will have calculated what his costs and his losses have been and will, quite understandably, try to levy them all against the unresolved issues that have been referred to arbitration. The respondent will, quite properly, disagree and the sides will become entrenched. Sadly, a dispute that could and should have been settled quickly at the time would thus be more likely to run indefinitely.

This brings us to what is perhaps the most significant factor to be considered when resolving any dispute, namely the cost of the dispute resolution process chosen. An early solution, reached by amicable negotiation following the proper control and management of the conflict, will save all of the parties involved from prohibitive dispute costs. Once outside experts or consultants are employed, costs inevitably rise dramatically. Expert consultants are hugely experienced and highly qualified, and as a result they charge very high hourly rates and, in addition, they will require reimbursement of their reasonable expenses. These consultants will need to travel and they may incur hotel bills. They will undoubtedly eat, charging you for their meals.

As well as these items, there are a host of other costs, which can be legitimately charged to your account. Over a prolonged period of time, weeks or perhaps even months, this will add up to a substantial amount of money. It is not unusual in the current economic climate to pay between £200 and £350 per hour for an expert or consultant and the hourly rates for solicitors and barristers will be higher still. Construction solicitors and barristers are highly trained and are probably the highest paid experts in the industry. The cost of involving them in a conflict or dispute can run into millions of pounds, especially if the dispute is destined to be a long one. There have been innumerable cases where the claim was successful, but the legal costs have been so great that they made the award look paltry by comparison.

Some time ago I was involved in a dispute, which arose because the two parties chose to defer the arbitration of major issues until the project completion. The client had caused the Contractor to carry out additional design work, install additional materials and, as a result, complete the works much later than programmed. Naturally the Contractor wished to seek a remedy for these breaches and as such he raised a claim for the sums he believed he was owed. The claimant's claims consultant did not stint himself and the 'loss' was valued at around £14 million. As is often the case in construction arbitrations, the proper value of the loss was much less, around £7 million to £8 million, and when the award was eventually made for a sum somewhat lower than even this, the arbitrator did not award the claimants their full costs. In this particular instance, which is by no means rare, the claimant paid his legal and expert fees and received a net recovery of only £2.8 million. Even the most meagrely talented negotiator would have been likely to recover £2.8 million against a reasonably evidenced £7 million claim, had the conflict been managed correctly from the outset.

Of course, there are times when the opposition simply refuse to negotiate or to discuss these matters sensibly or at all, and then litigation or arbitration becomes inevitable. Experience shows that in most instances a negotiated settlement is possible if both sides are prepared to be realistic about the outcome achievable in arbitration or litigation.

In addition to the more obvious costs discussed above, everyone in the industry can expect to suffer from the 'hidden' costs of construction conflict. Today our free enterprise system demands earnest competition in all sections of the commercial world, and construction professionals are not exempt from this crusade. Architects, engineers and quantity surveyors all face strong, and sometimes ruinous, fee competition when competing for new work. Once the work has been won, the professionals will try to complete the commission in the shortest time possible, always endeavouring to provide the client with a full and proper service.

However, when conflicts arise, they hungrily consume the professional's valuable and limited time. On a conflictive contract, letters, meetings, instructions and extra directions will abound and administration will be more intense. Partners, directors and higher level staff become involved as matters become more complex, and the fixed fee will become hopelessly inadequate to cover the extra expenditure of time. The resultant loss of revenue will become an internal accounting figure and, as such, will not impact on the project value as the professionals inadvertently subsidise the conflict.

The Contractor too can expect to incur additional costs on a conflictive project. He will inevitably have to bear a proportion of the money expended on additional management, staff and administration, as well as any disrupted site working. Obviously, the Contractor will try to recover as much money as possible, but even the most successful claim will be unlikely to recover more than the Contractor's normal or tendered overheads. The stricken Contractor will probably have to fund the excessive overhead expenditure, arising from on-site conflict, himself.

The client does not escape the financial penalty of hidden costs either. Whilst he has the benefit of an unwitting subsidy from both the Contractor and his professional team, the benefit does not result in added value or an improvement in his project. It is frittered away on wasteful side issues. In addition, the client will also have to bear some of the costs of conflict by reimbursing the Contractor for disrupted working and so forth. Ultimately, the unhappy client may find himself paying more and waiting longer for a project that cost more to build than it could hope to recoup in the property market.

It has been estimated that the costs of conflict could represent as much as 20% of the contract value on a contentious project. This can rarely be proven, as the hidden costs of conflict remain unrecorded.

Why, then, do we need to manage conflict? In my view, the answers are clearly stated above. Firstly, we need to avoid the unnecessary escalation of a minor dispute into a major conflict. Secondly, we need to control all conflicts to prevent a diminution of our good relationships with our customers and future clients. Thirdly, we must try to prevent the conflict from impacting on the timing and quality of the finished project. Finally, if these reasons are insufficient, then we should bear in mind that to resolve a construction conflict by legal, arbitral or even mediated means, will be significantly more expensive than effectively managing the conflict in the first place.

Whilst it may be thought that having a good conflict management team, who are well trained and who enjoy good interpersonal skills, is too costly, it is not expensive when compared with the alternative. The relatively small cost of training such a team will be offset very quickly against any litigation or arbitration costs, which would arise without their assistance.

Less tangible benefits will also flow as clients, who generally abhor conflict, remain faithful to you and continue to bless you with their patronage. You may also find that your relationships with those who work for you will be improved and that in the future people working together in this conflict-managed environment will be more contented.

When examined in detail, it becomes clear that the disadvantages of conflict will generally far outweigh the perceived advantages. So, as we now observe a need to address conflict correctly, we need to begin by examining the causes that are likely to lead to those disputes prevalent in the construction industry.

3 Causes of Conflict

As we have discussed previously, there are different types of conflict and, whilst generalisation is not always recommended in the conflict arena, we can categorise conflict into three main headings:

1. A conflict of ideas, beliefs or interests;
2. Conflict within an individual;
3. Conflicting personalities.

Each of these categories probably warrants its own chapter, because each category has different underlying causes. Unfortunately, life is rarely simple and a conflict that begins in one category may well migrate into others. However, if we are to comprehend the principles underlying the causes of conflict, we will do so more readily if we deal with them one at a time. The first section in this chapter looks at how a conflict of ideas can arise.

3.1 A conflict of ideas

Every human being is unique. We all share a common heritage and similar features, depending on our culture and race. However, beyond the obvious physical similarities, we are all very different from one another. Even identical twins, no matter how close, will have different thoughts, different ideas and, perhaps, different beliefs. It is argued, somewhat simplistically, that we are a product of our environment, our education and our experiences.

3.1.1 Environment

Our environment often affects who we become. People in different parts of the world become accustomed to their climate and may find other climatic environments exhilarating or oppressive, to varying degrees. Leaving aside

Conflicts in Construction: Avoiding, Managing, Resolving, Second Edition. Jeffery Whitfield.
© 2012 John Wiley & Sons, Ltd. Published 2012 by John Wiley & Sons, Ltd.

language differences, even people within the same country sharing the same language can have widely varying dialects. If the way we speak and look can be affected simply by geographical heritage, then it is clearly a possibility that our immediate environment will have a profound impact on our holistic human development.

There is a well-known proverb that states: 'The hand that rocks the cradle rules the world.' Our earliest development is supervised and guided by our mothers and fathers. The things that we hear as children are often immediately accepted as fact, no matter how unfounded they may be in reality. I read recently of a business philosopher whose mother had a favourite expression: 'The black cat is always the last one off the fence.' It seems that as a child he simply accepted this saying without really understanding its meaning. Eventually, having realised that he was repeating it as an adult, he quizzed his mother about the saying, who would only say, 'It's true, I've seen it with my own eyes,' before refusing to discuss it any further, as mothers sometimes do.

Just as sayings, dialects and accents can be picked up from our neighbourhoods, so can our attitude to authority, our view of society and our ability to see the future with hope and optimism.

As much as we would like everyone to be treated equally in society, some children are obviously more materially privileged than others, some are more emotionally privileged and some are, inevitably, more intellectually able. It has been argued for generations that our *ideas, beliefs* and *interests* are formed to a large extent by our environment, our childhood and our education.

For example, before legislation outlawed the hunting of foxes with hounds, the son of a foxhunter may have wanted to perpetuate the local hunt for any one, or a mixture, of these three separate reasons:

1. He may have thought that foxhunting was the natural thing to do – an *idea* engendered by being raised in the countryside.
2. He may have believed that foxhunting was the best way to control foxes – a *belief* arising from anecdotal evidence provided by his family and friends.
3. He may have considered foxhunting to be cruel, but he may have continued because he feared negative family reaction, the loss of affection or even the loss of an inheritance – an *interest*.

As a child, I found an old-fashioned toy called a kaleidoscope in a cupboard. It was a cardboard tube with pieces of coloured glass held captive at the bottom in mirrored, translucent plastic that transmitted light. At the top was a viewing lens. When I shook the tube, the glass would move and every time I looked inside, the light shining through the bottom of the tube would reveal a different arrangement of shapes and colours. Try as I might, I was unable to reproduce exactly the same effect again, because there were too many variables involved. So it is with people and their environments. No individual, once conceived, delivered and parented, will ever exist in exactly the same form again.

3.1.2   Education

Man differs from other life forms in many ways. One of the most significant ways in which we differ, even from primates, is in our ability to reason. Animals and humans both have instinctive reactions that protect us from danger or injury, but modern man has developed this instinct into a reasoning process that enables us to make far more complex decisions than those simply affecting our own well-being. The process that moves us from instinctive reaction to creative or reasoned behaviour is called education.

The education of any individual will be a mixture of both structured and unstructured learning. Structured learning is what we generally have in mind when we use the word *education*. To improve our education, we attend school, colleges and universities to add intellectual knowledge to our innate intelligence. The curriculum is designed to build understanding block by block. Unstructured learning is often accumulated much more quickly, because it is interest led. By a series of simple experiments, children can quickly find the answers to questions such as: 'What does soap taste like?' 'How much peanut butter does it take to coat a DVD?'

As youngsters, we observe others and listen to their voices and we learn to walk and to talk. The greater our learning, structured and unstructured, the greater will be our ability to make reasoned decisions. In short, the wisest choices are made with full knowledge and a rational approach to decision making. Both of these aptitudes are learned aptitudes acquired by an educative process.

Without education, our ability to reason and create ideas of our own would be very limited and we would find it extremely difficult to communicate our ideas to others.

3.1.3   Experiences

Environment and education can often be almost identical from one child to another – twins, for example. So, apart from biological factors, what other factors will determine why these twins often react differently when faced with an identical problem? There are many factors, but perhaps the most significant is their treasure chest of experience. One will have tried similar things before and succeeded; building their self-confidence and setting them free to move forward to other challenges. Perhaps the other twin will have tried previously and failed and, having felt the pain of failure, will be reluctant to make bold moves forward again.

However, failure need not prevent us from making future attempts to succeed. Optimistic individuals may have tried and failed in the past, but they recognise that failure often precedes great success and they may even be excited about trying again. In essence, our experiences can either be limiting or liberating; it largely depends on our outlook. Psychologists tell us that every experience, good and bad, is stored in our memory and will be accessed subconsciously when we need to make decisions. To prevent our subconscious minds from dominating our decision-making processes and limiting our progress by reminding us of the pain of failure, we need to review our past errors and use them as learning experiences.

Thus it can be seen that our past experiences may have a significant impact both on the number of ideas we have and on how daring those ideas are.

3.1.4   Ideas

We can now begin to understand why it is that when we set a dozen people the same task, they each address it in a different way. The surprising thing is that many of these individuals will produce a perfectly acceptable end product, despite the apparently unorthodox way they have approached it.

These individualistic ideas of how to approach a problem can cause a conflict if:

1. One side *knows* it will not work.
    That is, available research, evidence or personal experience shows conclusively that the idea is doomed to failure.
2. One side *believes* it will not work.
    That is, available research, evidence or personal experience suggests that the idea is unlikely to succeed.
3. One side believes that *their idea is superior*/will work better/will favour them.
    Past experience has led to a conviction that this idea is more likely to produce the desired outcome.
4. One side thinks that the idea may work but is afraid because it was *not their idea*.
    A reluctance to acknowledge that another person has found a possible solution that has eluded you, or a fear that the other person is superior in some way, may lead to a reduction of self-esteem in fragile egos.
5. One side is afraid that the idea will work but *will benefit only the other side*.
    Not really a conflict of ideas but a conflict in outcomes, discussed later in this chapter.
6. One side believes it may work but that it is *too impractical, outlandish* or *hazardous*.
    Past experience suggests that the idea will work, but that it may be too costly in financial, safety or interpersonal terms.

Conflicts of ideas are important, because they arise frequently in construction and are closely related to conflicting beliefs.

3.2   A conflict of beliefs

If all human beings could discern absolute truth, without fail, then there would be no scope for belief, no need of faith in our fellow man. The number of different religions in the world and their disparate creeds tells us, if we really need to be told, that man is fallible. Intelligent and reasonable men are clearly capable of believing falsehoods. Human beings can be misled by lies and deceived by tricks. Because the truth is not always obvious, we need to rely on reasonably held beliefs.

Belief usually arises from the forwarding of an idea and our acceptance of that unproven but feasible proposition. If we are exercising real belief, we will be confident that the proposition is correct – it becomes, in essence, our true opinion.

In construction, as in all other facets of life, there are uncertainties and so there is scope for genuinely differing beliefs. However, a belief must have some foundation. There must be some reason for a firm opinion stubbornly held.

Expert witnesses are allowed to give opinion evidence in a court of law, a luxury not afforded to witnesses of fact who can only testify to things they know. But evidence based on an expert's firm opinion will only be acceptable to the tribunal if the opinion is justified and supportable by a body of opinion in the expert's field. Outlandish or baseless opinions will be ignored, no matter how fervently they are held. Beliefs should be realistic, supportable and must not fly in the face of the facts.

The Construction Industry is filled with individuals who have differing beliefs and who are happy to promote their beliefs, views and opinions loudly. This, of course, leads to conflict where those beliefs do not concur with the beliefs of others. Initially a conflict of beliefs will appear to be similar to a conflict of ideas, but there is a difference. Unlike an idea, a belief or opinion must be credible enough to sustain the confidence of a reasonably objective person. Real conflicts of belief arise when:

a) One of two or more propositions could be true, but no evidence is *currently* to hand to prove which, if any, are true.
    We may have a conflict of beliefs where one person believes the contract allows for an extension of time for adverse weather, because he would never have accepted a contract that did not contain such a provision. He cannot prove his case instantly, however, because the contract documents are elsewhere.

b) One of two or more propositions may be true, but no actual evidence exists to prove which, if any, are true.
    In this case, a contractor may believe that his client has caused the site operatives to work unproductively by issuing late instructions. However, if a contractor has not kept records of just how productive his workforce was, or should have been, it is an unsubstantiated belief.

As will be observed from the above examples, some conflicts of belief can be resolved easily given time and evidence, whereas others will almost inevitably have to be compromised or adjudicated.

A conflict of ideas or beliefs will not usually become problematical, if those ideas or beliefs are not expressed. Even when they are expressed, a conflict will not become active unless we have a strong interest in our ideas or beliefs prevailing over those of others. Rightly or wrongly, we do have an interest in our beliefs being approved by others. The reason for this interest is simply that human beings, as a race, obtain their self-esteem from personal success, and success is often measured by the number of occasions on which our views override others' views or are accepted as correct by others. We will naturally have a keen interest in our own ideas prevailing, and others will have just as keen an interest in their ideas dominating. When these ideas, or elements within the ideas, are in opposition, there may be an active conflict, but this is not necessarily the case. Perhaps an example will explain the principle more clearly.

Tim fervently believes that the Labour Party would run the country better than the Conservatives. Tom, on the other hand, believes that the Conservatives would be better at running the country. The ideas and beliefs held by Tim and Tom, if unexpressed, cannot be known by the other side and so cannot cause conflict between them, even though the ideas themselves are in conflict. If these diverging ideas are expressed publicly, then there is a conflict between the two competing ideas and the individuals recognise that they are in conflict. However, even at this point in time, there may be no active conflict because the election is still years away, and so neither Tim nor Tom have an overriding or strong interest in imposing their views on the other (or the public) at this particular moment. If it was a General Election year, the picture might well be different, in that the opposing views have now been expressed and the candidates have an urgent interest in persuading floating voters that their views are correct.

In many cases, the urgent need for a decision acts as a catalyst to a conflict becoming active. Thus, there are three separate factors leading to an active conflict:

1. knowledge of the conflict;
2. an interest in promoting our position; and
3. an urgency of resolution.

3.3   A conflict of interests

In conflict terminology, the general term *a conflict of interests* would include a conflict of ideas and of beliefs, but for this book I consider it important that we examine each element separately and in depth. What most people usually describe as a conflict of interests is actually a one-sided interest in a particular outcome. So the third element that can give rise to a conflict of interests is a *conflict of interest in an outcome*.

Usually, when we have an opinion or an idea, that opinion or view may be reconsidered if further evidence is provided. For example, we may believe that the world is flat but, having heard the testimony of someone who has sailed around the world, we may be prepared to review our belief and change our publicly stated opinion. If, however, we have a vested interest in the outcome of such a conflict, for example we are the highly paid chairman of the Flat Earth Society, then we are far less likely to change our publicly stated beliefs so readily.

It is not unusual, therefore, to have a disagreement in which there is no conflict of beliefs but only a conflict of interests in the outcome. Where one man wishes to buy a car from another, they may both know that the car is worth £10 000, but the buyer only has £9000 to spend. The buyer consequently argues vehemently that the car is only worth £9000, to give him a chance of purchasing the car. In this example, there is no difference in ideas or belief but only a difference in interest in the outcome, and so it can be seen that a conflict can arise where there is no underlying disagreement at all. This is often the case in construction disagreements, where one party needs an improved financial settlement to satisfy its funders and so it creates an unsubstantiated claim in the hope that it will be paid in part. The problem that arises from an overriding interest in outcome is that the claimant's position cannot be changed by reasoning or the introduction of evidence to the contrary. I recall reading that a shrewd Victorian parliamentarian was quoted as saying, 'You may change my mind but you will never change my vote.'

The fact that a significant proportion of construction disputes arise from *A conflict*, as opposed to emanating from the urge *To conflict*, suggests that one or more parties may have a strong interest in a particular outcome. This type of interest will usually lead to position taking and to a lack of tolerance of opposing ideas or views.

By now we can all understand why the parties to a dispute may have a conflict of interests, but why are such conflicts so prevalent in the construction related industries? Why do we have so many differing ideas, beliefs and interests? Even our eminent judiciary have described the construction industry as being 'fertile ground for conflict', so what lies behind this Fertile Ground Theory?

3.4   The Fertile Ground Theory

If we exclude conflictive people from the equation (they are discussed in a later chapter), then we are left with a single major contributor to the Fertile Ground Theory. Most disputes in commerce generally, and in construction in particular, arise from some form of *uncertainty*. The disputes may well be disguised but all will involve either a conflict of ideas, of beliefs or of interests and, as we have already observed, conflicts of ideas and beliefs are most likely to arise where there is some uncertainty. Unfortunately, the construction industry is, by its very nature, beset with uncertainty. Because uncertainty is present in the construction process, aggressive individuals are able to capitalise on any ambiguity and manufacture a conflict where there was no conflict originally. The real reasons that the construction industry is rife with uncertainty include the following:

3.4.1 Prototype

Every new project is essentially a prototype. It is unique in its own way. It may have similarities to another project but many things will inevitably be different, including the team who build it. In the motor industry, prototypes are produced for each new model for a variety of reasons, for example:

- to see if it works as anticipated;
- to see what it looks like at life size;

- to decide how best to produce it;
- to determine whether it can be built economically;
- to identify opportunities for economies of scale, mass production, batch production;
- to bring the designers and producers together to build in quality and safety;
- to resolve problems before a whole labour force is disrupted.

Because each new construction project is arguably a prototype, many of the lessons that are learned from a prototype car must be learned during the actual construction process of a single project.

For example, a new building project with its unique ground conditions has never been attempted before and so we can only use our experience of comparable conditions to devise a construction methodology that will work. Others may have had different experiences and so they may validly hold a different belief and suddenly there is uncertainty as to the best method of foundation construction.

Construction problems have to be resolved as they arise, often as the site labour force waits for instructions. There will be a number of possible solutions, but the most economical will not necessarily be the most convenient, or the quickest. So immediately we introduce uncertainty as to time and cost.

Two-dimensional drawings are simply incapable of communicating the scale of the completed, full-sized, three-dimensional building to an inexperienced viewer. To the client, the finished project may seem quite different from his original expectations, introducing uncertainty about the actual aesthetics of the project in its built environment.

Finally, there is the introduction of new construction products. Offshore wind farms of 1000 plus turbines are being planned and executed around the coast of the UK, yet no-one has any experience of bringing so many turbines, of the size and complexity of these new models, together in such large numbers. Budgets and programmes for the completion of such projects are best estimates, but this lack of historical data leaves scope for one party or another to argue that the works could have been executed more quickly, more cheaply or more safely.

3.4.2 Change

Everyone who has ever been involved in a construction project will know that change is inevitable. In many, if not most, instances, the work on site commences before the design is finalised. This acceptance of change is widely accepted in the industry and fast track projects are often deliberately planned to allow design on the later elements to continue as the earlier elements are constructed. This causes uncertainty as to what the finished project will include and confusion as to what the various contractors were expected to include in their prices.

3.4.3 Delay

Most engineering and construction contracts anticipate failure. They are written in the expectation of one party's inability to comply with their bargain. Clauses are included, which allow the Architect to provide late information, allow engineers to issue changes to the contracted works, and anticipate contractors will complete their works late owing to their own shortcomings. Given all of these expectations, is it at all surprising that the parties involved often live up to them and bring in a project late and over budget?

What greater uncertainty could we build into a contract for a new project? If uncertainty is the main criteria for determining fertile ground, then it is indeed true to say that construction is a fertile ground for conflict.

If uncertainty gives us the fertile ground, then a difference in ideas, beliefs or interests will give us the seed. Even so, we need to water and nurture the seed for it to grow. A conflict is of itself inert. For example, a conflict of beliefs over what the wording JCT Clause 26 actually means is simply an intellectual or semantic debate without some catalyst. Unless there is a delay to the progress of the contract, Clause 26 will remain unused, and a disagreement over its meaning will be unimportant. Should there be a delay, however, and it is found that the wording of the clause impacts on the possible entitlement of either party, then the conflict is about to crystallise.

If a conflict is to escalate into contention, then someone must champion one view or belief against another. We need a dispute to nurture the seed. In construction, three things cause most, if not all, of the disputes – Quality, Time and Money.

3.4.4 Quality

Unless quality is adequately defined, it remains a subjective issue. A high quality of finish will mean different things to a plasterer and to an architect, but an architect may try to impose his understanding of the definition on the plasterer. Objective standards must be sought to define materials and workmanship, if there is to be a reduction in avoidable conflict. Such objective standards can be provided by citing the British Standards or by naming a unique product.

We also need precision and accuracy in our description of what is required. If we do not make it clear, then we leave the clarification of our intention to others, who will interpret it to suit their needs, not ours. If we fail to tell people what we want to buy, then they will provide us with what they wish to sell. That is rarely satisfactory and often causes conflict.

Our definitions of quality should be realistic, and appropriate to the price we are prepared to pay. It is sometimes the case that the client will specify a much higher standard than he really requires, only to balk at the price once the money has been spent on his behalf.

Briefs to designers should concentrate on the items the client finds important. If they do not, then the Architect will tend to give priority to aesthetics and the Engineer to structural stability. The completed building may then either be short of space and look marvellous, or withstand an earthquake and look awful.

Quality and finish are very much in the eye of the beholder, and failure to adequately specify your requirements will lead to a conflict of beliefs, one where the Contractor really believes the quality is acceptable, or a conflict of interests, where the Contractor simply provides the lowest cost solution.

3.4.5 Time

It is rare to see a dispute arise from a contractor having too much time, and so we will concentrate on the real issue, which is too little time. Time is only important in construction, because we set completion dates *and* provide for the remedy of liquidated damages if the date is not met. Almost everyone directly involved in the industry has a profit motive. We know that time is money and so to deprive someone of time can be as galling to them as robbing them of their money.

When specifying the time for the completion of a project, we need to be realistic. We need to anticipate those problems that will consume valuable time and resources. For example, allowing the Quantity Surveyor less time to prepare his contract documents because the design was late does not save time, it merely increases the prospect of error, which will add more time, and expense, to the overall process. The whole constructional process needs to be time controlled from inception to occupation. To finitely monitor the contract works, with a network accurate to one day, is palpably useless if the design information progress is monitored weekly or not at all. Each element of the construction process must have an appropriate and agreed timescale, independent of the shortcomings of others' progress. To shorten future activity durations, because early milestones were not met, is not the answer. Such a remedy will only move the eventual dispute back in time.

3.4.6 Money

Money may make the world go around but it brings more construction ventures to a halt than any other topic. If the price for a project is wrong, then we can expect problems and conflicts to harass that project throughout its life. As with time, the money must be adequate for the project in hand. If fee competition is too harsh and an architect accepts a commission on ruinous terms, then he will not be the only sufferer from the effects of his foolhardiness. He will try to design the project with lower quality staff and in fewer hours; he could skimp on his overseeing and quality control roles; he might even decide to cease practising altogether and leave the client with a time and money problem far greater than any saving in fees. Adequacy of funding and appropriateness of funding will reduce the inevitable conflicts that arise over financial matters.

Money is inextricably linked to the quantity of work to be undertaken. If we deliberately or mistakenly mislead a contractor into believing that ground conditions are better than they are, he may well submit a lower price, but the matter will not end there. We may even have him firmly bound into a cleverly worded contract, which seems to remove any remedy he may have, but no contract can make someone do what they do not want to do. In one way or another, the Contractor will ensure that he recovers his losses and, in doing so, any feeling of responsibility for a job well done will vanish.

Honesty and fair-mindedness in the preparation of contract terms is essential to avoid conflict. No-one will win a war of attrition fought on a long-running project; there can only be losers. If only one area of conflict can be eliminated, then this should be it, because it impacts in such a damaging way on every aspect of the project.

Once we have established that 'a conflict' can arise in one of many different ways, then we need to consider the people involved in a conflict, as their response and reaction to the conflict will govern its impact on the project.

3.5   Internal conflict

Having discussed in some detail the reasons behind a conflict, we must move on to examine the impact of conflict on the people who are involved in conflict. Relationships between people can govern the length and the depth of any dispute arising from a conflict, but before we discuss this, we need to examine the internal reaction to conflict within a human being. This internal aspect may also have a significant impact on the way a conflict will develop.

Internal conflict differs from external conflict, in that it deals with opposing positions within a single individual. When an individual is faced with a choice, what will he decide? This, in essence, is internal conflict.

Psychologists have a series of theories on internal conflict, which emanate from the Avoid–Approach Principle. This principle suggests that whenever a person has to make a choice, he will seek the most pleasant outcome for himself. For example, given a choice between a piece of chocolate cake or a slap in the face, the majority of people would choose the piece of chocolate cake. From choice, we would naturally wish to avoid the slap in the face and instead approach the chocolate cake. This is an example of an avoid–approach technique. When, as in this case, the choice is simple and clear-cut, we have a very easy decision to make – to approach one good or worthwhile cause and to avoid one less worthwhile, unpleasant or painful option. There is no real conflict in this decision-making process.

The real internal conflict appears when we have two equally attractive options. For instance, the choice may be between the piece of chocolate cake and a slice of strawberry cheesecake. We may only choose one. We now find ourselves in an approach–approach situation. We would like to approach the chocolate cake, but we would like to approach the cheesecake too. The decision is now quite different. We have to decide between the two equally attractive options, and this causes conflict within us.

If we can have two pleasant options to choose from, we can also have two unpleasant or painful options. Perhaps the choice now is between a slap on the left cheek and a slap on the right cheek? We now have an 'avoid–avoid' situation. Whichever we choose will be unpleasant, yet we must choose one, so the decision must he based on our appraisal of the particular 'avoid–avoid' question we currently face. Most people would have few problems in making this decision, however, because we realise that one option is no better or worse than the other. They are of the same value, which makes the decision an easy one, in theory.

A further problem that we may face is the single subject avoid–approach scenario. Decision making in these cases may be more difficult, because a number of different factors will complicate the process. Returning to the example of the chocolate cake, perhaps you would like to take a piece. However, at the same time, you are diabetic, and know that this would cause problems, so you should avoid it. This simple example illustrates an avoid–approach conflict.

Clearly, when faced with two pleasant alternatives, we try to judge which alternative is likely to be the most advantageous option and we base our decision on that criterion. Similarly, when faced with a decision between two options with negative consequences, we try to look objectively at which is the lesser of the two evils. In the example given above, we have to decide, for instance, which alternative is preferable – the pleasure obtainable from eating the chocolate cake, with its possible detrimental effect to our blood sugar levels, or avoidance of the cake altogether, which means missing out on both the pleasure and the ill effects.

The decision-making process is complicated further by the double approach–avoid situation. In this example, we are faced with a choice between a carrot and a slice of chocolate cake. We would like to approach the chocolate cake because it tastes so good, but we would also like to avoid the calories it provides. On the other hand, we would like to approach the carrot because it is good for us and has fewer calories but, when all is said and done, it tastes like a carrot.

Internal conflict is often much deeper even than this. Another situation we often face is where we have a conflict because we have to choose between several options that are priority-based rather than simple decisions as to which route is best. You may be unable to pursue the option you would like, because other less attractive choices must be considered first. For example, you receive a sum of money. You would like to spend it on a holiday, or some new clothes, but you have debts to pay and bills to settle, so a decision based on priority rather than desire should be made. Often these priority-based decisions have to be reached very quickly, because no time is available to make a logical and reasoned examination of which to approach first.

Let us suppose, for example, that you are in the kitchen. The microwave oven beeps, a pan of potatoes is about to boil over, the doorbell rings and so, too, does the telephone. Each item requires immediate attention. There is no time to decide which, if any, is the most important. An instant decision must be made, even if it proves to be wrong. The alternative is to do nothing, which is the worst possible thing you could do. Such situations are often the most difficult to cope with.

The choices that we have mentioned thus far are, of course, relatively simple and of little consequence, but we do have much weightier decisions to make in our lives, which can have far-reaching results. Generally speaking, the decisions involving the greatest consequences also cause within us the greatest conflict. Even so, difficult decisions can be made by objectively studying the pros and cons, and by doing a cost benefit analysis to identify which option is the most suitable for us in a given set of circumstances and at that particular time. We should bear in mind that in different circumstances, or at a later point in time, the most desirable option may well be different, so our choices must always be made on the basis of all of the objective criteria. These can be described thus:

- Which is the better option?
- Which is the better option at this point in time?
- Which is likely to be the better option in the future?
- Which is the better option in differing circumstances?
- What are the foreseeable consequences of each option?
- Is the decision based on preference or priority?
- Does the best option conflict with our internal value system?

Having considered each step, we can proceed to make the decision. We will often do this instantly, not realising that we have progressed through these steps, because most decisions are made automatically or instinctively.

How, then, does this internal conflict impact on the construction industry? All construction professionals and workers should have the capacity to reason soundly and make decisions based on objective criteria. The problems begin to surface, however, when decisions must be made on subjective criteria, or when a choice must be made as to what is right or wrong in a moral or ethical sense. To illustrate this, we return once more to our original dilemma of choosing between a piece of chocolate cake or a slap in the face. We can complicate the decision by introducing a moral question: Choosing the chocolate cake would, in fact, mean stealing it from someone else. The choice now is that we either steal the chocolate cake, or we will be slapped in the face. The question now becomes a very different one, and more difficult to resolve.

Each individual has an internal value system, which is comprised of our ethics and morality. This value system will differ slightly from one individual to the next. As discussed in the previous chapter, it will be affected by our environment, education, experience and many other widely-differing criteria. Within the construction industry, we have further group value systems, such as codes of practice for professionals including quantity surveyors, architects and engineers.

In addition to these, we also have peer group pressure. This imposes yet another set of values upon us – the values of the peer group itself. All of these are important, yet there are still more values imposed upon us. These imposed values are the laws of the land. Many of these laws will coincide with our internal value system, our ethics and morality, but some do not. These laws are imposed not by our value system but by external forces, in this case by the authority of the legislature. It is important for us to be able to separate each different component of our value system, so that we do not become confused when faced with an ethical problem in the future.

We have discussed so far that internal conflict is basically a question of making choices or decisions based on given criteria. We should also understand that those decisions, which are made objectively, should actually be relatively easy to make. In addition, we have examined the impact of a value system on our decision-making process. We now need to put these things together, to discuss what kind of difference our value system can make when we have to choose an option or make a decision.

Values are unusual in many ways, because they are an inherent part of our being and, unless we examine the reasons why these values are held, we may not know their exact origin. It is clear that our values are tied almost inextricably to our personal ethos – the way we grew up, the way in which we deal with others, and the way we relate to ourselves. Our personal pride, our self-esteem and our self-worth can all be affected by our internal value system.

Of all the conflicts an individual may come up against, the most difficult to overcome will almost certainly be those conflicts pertaining to the internal values held by the individual. It may be possible to persuade someone to accept a reduction in salary for a reason perceived as being a good one. It may also be possible to persuade someone to accept your views on politics, for instance – because they do not have to change their own values to do so. However, it will be extremely difficult and, in most cases, impossible to persuade another to steal, kill or commit some other crime because their value system would prevent them from doing so. If a person is persuaded to breach their value system by doing something with which they are instinctively unhappy, then that individual may be psychologically damaged. They may lose their self-esteem, their self-worth and sometimes even the will to live. A well-developed internal value system is fundamental to a person's well-being and ultimately to how successful that person will be in every aspect of his life.

A well-known prostitute wrote a book about her experiences, in which she explained her first venture into prostitution very poignantly. Times were hard where she lived. Her family were poor, and hungry. She was the only one capable of earning any money. She would not normally have considered prostitution as a means of earning a living, but she finally decided, in desperation, that she would do this just once, as they needed money for food. It was an objective decision, carefully made.

However, having done this just once, having acted against her internal value system, she felt so awful, so dejected, so used and abused, that she lost all of her self-esteem and self-worth. Soon her internal value system had been almost totally suppressed with regard to moral issues. She was now in a situation where moral or ethical questions about prostitution were no longer a barrier to be overcome, so the decision to go out and continue to earn money in this way became a simple one. This is an extreme example, but it illustrates the dangers of going beyond one's internal value system.

For most of us, our value system will not disappear all at once, as in the above example, but it can be eroded a little at a time without this being noticed initially. Perhaps we begin this process by rationalising that there is no real harm done by taking a pencil we really need from a shop display. By a process of degrees we may move on to steal items of increasing value. Eventually we arrive at a point where our value system, or conscience, no longer intercedes when we steal, because that particular moral value has been subdued. Oddly enough, the balance of your value system will often remain in place as far as other moral or ethical issues are concerned. We can see from these examples that our internal value system is very complex. It is because this system differs from one person to another, that other human beings become hard to understand.

Another issue to be considered is the subject of self-justification. Occasionally, we all do things that we know instinctively are wrong and are in conflict with our basic value system, yet we justify the reasoning behind our actions. We try to provide ourselves with a convincing reason as to why we have acted in this way, and in so doing we are deceiving ourselves. Perhaps we decide to avoid the tax due on a cheque, which arrives from a client in the post, by banking it directly into our personal account. We know that this is wrong, and that it goes against our moral codes, but we rationalise that the taxman already takes sufficient of our income, and that no-one will ever find out anyway.

This rationalisation process, however, usually only works as far as relatively minor transgressions are concerned. If we were to commit a major crime, we would not be able to avoid the guilt that our rationalisation tries to salve. There are those who make an unwise decision and try to rationalise it away, only to find that the unwise decision niggles at them for years. They find that they are unable to forgive themselves, because they realise that they have breached their own value system. If, in addition, they have failed to put right their mistake, the feelings of guilt will be intensified. Often it requires an admission of the mistake, followed by restitution where possible, to make such a person feel better about themselves.

As an example of this, we can look at the Inland Revenue example cited earlier. A friend of mine, who was a Tax Inspector, said that over the years tax collectors have become accustomed to receiving unexpected sums of money from taxpayers who know they have underpaid tax. This type of restitution will salve the consciences of such people, because they have a need to feel that they have done the right thing, even if it has been left very late. To resolve such a matter of conscience at so late a stage is, of course, preferable to leaving the matter unresolved, from the point of view of all parties concerned.

Honesty, fairness and integrity in all things are of paramount importance. It is important for any individual to keep his value system intact. Breaching that value system and leaving the matter unresolved will bring us weeks, months or perhaps years of feeling uncomfortable and unhappy with ourselves. We will suffer feelings of reduced self-worth and low self-esteem, all of which could be avoided by being honest and living within our personal moral and ethical value based system.

3.6   Internal conflict in construction

Most individuals working within the construction industry are honourable, fair and honest, wishing to carry out their duties in an ethical and morally acceptable manner. They will usually be upright and reasonable in the decisions and choices they have to make. They will also be men and women who live full lives outside their construction jobs. These people need to feel comfortable by applying the same values at home and at work. They can then live contentedly with their partners and families, as well as with their occupation within the industry. The majority of people in construction recognise this fact. There are a few, however, who believe it is possible to live a scurrilous life within one's occupation, with no reflection upon one's family or outside interests. Of course, this is simply not the case. It would be a very unusual person indeed who could have two separate value systems, and live each successfully, without some kind of psychological trauma, or subsequent negative mental effects. A common term used to describe people unaffected by their behaviour towards others is a sociopath.

To enjoy real peace of mind, each of us needs to observe our internal value system in all aspects of our lives, making all judgments and decisions in accordance with it. Lies told, deceits practised and rationalisations made will all remain in our minds if left unresolved. Truth, honesty and integrity set us free, enabling us to make real personal progress. Those individuals who find themselves held back from such progress, or who suffer mental dilemmas, often have within their minds unresolved internal conflicts. We should therefore recognise that we can move forward only when we are able to accept and resolve our past mistakes and failures. We need to make a commitment to ourselves that we will do better in the future, having learned from our errors. Those who do this will feel comfortable with their lives and will consequently be better able to interrelate with others.

If we find ourselves feeling uncomfortable with whom we are or what we have done, we will be less able to deal successfully with other people. A feeling of inadequacy or of low self-esteem will inevitably impact on our relationships with others. If a person feels as though he is torn apart internally because he has betrayed his value system, he may subsequently make everyone around him as miserable as himself, even if this happens subconsciously.

If all our decisions in life were simply between misery and joy, our lives would indeed be much simpler. Life, however, is not simply black and white. There are innumerable shades of grey in between, and the decisions we make require careful consideration. Sometimes short-term benefits can result from dishonesty and lack of integrity. It is possible, therefore, to breach the value system, to be dishonest, and to 'get away with it'. These short-term benefits may thus be construed as achieving false happiness, and it is sometimes difficult to differentiate between what is false and what is real. Thus, our decision-making processes can be complicated by what appears to be success in the short term, but real happiness and success can only be achieved by having made the right decisions.

Those people who attempt to lead duplicitous lives are often found out by others long before they themselves realise that their way of life has been discovered. This will damage their relationships with others, both socially and professionally, and will ultimately lead to damage to their own self-image. Most of us know that when we try to be dishonest with ourselves, it does not work for very long. The same is true of when we try to be dishonest with others. The man we criticised yesterday will eventually hear the criticism. The man from whom we steal will ultimately notice his loss. The man we may try to persuade improperly to act against his own value system will, in time, recognise his mistake. In each case, our relationship with that person will be damaged, in many cases beyond repair. Our reputation, or our good name, will also be damaged as news of our deceit and double dealing spreads to others who know of us. With our reputation in question, our lives could soon become empty, and our subsequent lack of self-worth may lead to severe internal distress or depression.

The honest man needs only to remember the truth. The dishonest person needs to memorise a string of lies which, if he fails to memorise them correctly, can cause him to lose friends and influence. Even if he memorises his story convincingly, he will put increasing pressure upon himself, because each time he lies or deceives to perpetuate the story, a little more damage is done and more stress, more pressure and more debilitation of the mind occurs.

As we discussed earlier in this book, the construction industry is highly motivated by the accrual or loss of money. This factor will persuade some people to override their own value system, bringing them into conflict with their otherwise strongly held moral and ethical beliefs. As we have seen, these beliefs are gained in a number of ways and include:

1. *The internal value system*: This is built up over a period of years, influenced by the examples shown by parents, friends, peer groups, education, environment, religious beliefs and personal experience.
2. *The value system of the social/peer group*: The people with whom we mix and spend our time will have a given set of values as a group. Some of these values may include loyalty to the group, following club rules, etc., but do not usually conflict with our internal values, such as being honest, faithful family men who respect the environment and obey the laws of the land. In some extreme cases, gang membership for example, the desire to be accepted by the group may be such that we would lower our standards in order to fit in.

3. *Professional values*: These values are additional laws, rules and regulations imposed upon us by professional bodies.
4. *The laws of the land*: These laws impose restrictions on our actions to a large degree, inasmuch as they have an impact on our fellow citizens. They are necessary, however, in order to ensure peace and security.
5. *Religious or other moral–ethical standards*: These are a series of rules and instructions, which are governed by the churches, by our mentors, or by our role models. They affect our ability to have total freedom in our actions. A subject which may be in accordance with the laws of the land may still be regarded by us as being morally wrong.

Now let us examine how each of these types of values or laws impact on us in our role within construction.

Firstly, we need to examine our basic internal value system, which has been built, block by block, from our childhood by experience, education and parental control. This very basic value system will give us rules by which to live. The rules are a controlling influence on the otherwise natural man, which anthropologists observe in some primitive societies where these types of rules are not present.

These rules, or moral obligations, include such things as:

- We should not hurt other people.
- We need to share.
- We need to say sorry when we have done something wrong.
- We need to love other people.
- We need to care for those who are less fortunate than ourselves.
- We should be content with what we have rather than be greedy.
- We should not take what does not belong to us.
- We should not tell lies.
- People are always more important than things.

Of course, there are countless others, and all will contribute towards building our own personal value system in its basic form. Other blocks are continually being added, and the system builds itself until our value system is fine-tuned and ready for each individual to use in relation to his role in society.

Secondly, we are undoubtedly influenced by our peer group. These are people of perhaps a similar age, perhaps our colleagues, and certainly our neighbours in the community in which we live. This group will probably include our friends, perhaps relatives and colleagues as well. If the group to which we belong has higher standards than we do, we will have a tendency to improve our own standards to match theirs, so that we can gain acceptance within that group. However, it is sometimes the case that the group's standards or values are lower than ours, and we may find ourselves so desperate to be liked, wanted or accepted by the group, that we reduce our values and lower our standards of integrity and honesty in order to match theirs. We all have a natural desire to be wanted or accepted and so we must beware of the type of groups with which we associate. The types of values that might be reflected in a peer group are such things as:

- Be considerate of your friends.
- Never steal from a friend.
- Be available for your colleagues.
- Be trustworthy.
- Comply with the basic rules and regulations of your location, environment or workplace.

Alternatively, we often experience negative peer pressure, which attempts to impose a different set of standards:

It's okay to drive after a couple of drinks.
The taxman will never miss it/never find out.
If you get him/her drunk, he/she will agree to anything.
Taking a day off sick when you are not really ill is no more than you are entitled to.

The type of peer group pressure to which we are exposed will enhance or detract from our internal value system. If we become desperate enough to fit in with a particular group, or to meet another individual's aspirations, then we may find ourselves with a value system that is more akin to theirs than our own.

Thirdly, let us look at the laws of the land. Whilst the various rules and regulations imposed by the State do not always impact on our internal value system, because we may consider some of them unnecessary, irrelevant to us, or even unwise, many of the laws of the land will reflect upon our decision making internally. In reality, we all accept that there is a need for control, for rules and regulations, for restrictions against certain types of behaviour. For example, we accept that if there were not anti-litter laws, the world would soon be awash with litter and debris left behind by careless passers-by or thrown from cars. Life would be much less pleasant and there would be damage to the environment, and so most of us believe that these rules are relevant to us. Even the majority of those who break these rules recognise the validity of them.

Rules and regulations will eventually impact on us internally, inasmuch as our internal value system should tell us, before the law of the land does, that we should not do something, such as drop litter, rob banks and so forth. Of course, there is a far greater impact on our lives and on our value system should we choose to break these laws. We may be punished for those laws we breach, and this punishment may include financial penalties and loss of freedom. This threat of punishment may help us decide more easily whether or not we wish to break that law.

The laws of the land are not the only laws for which penalties exist for those who break them. Our peer group has the power to ostracise us, for example, if we act against the group value system. Professional bodies, too, have the power to punish certain misdemeanours, and they are discussed below.

Professional bodies were originally established to monitor the education and professional behaviour of their members. These groups have become very powerful, both inside and outside the construction industry, and have considerable influence on the professional lives of their members. Designatory letters after professionals' names certify not only that they have passed the required examinations, which make them suitable for a certain level of employment, but also that they observe the aims and objectives of the body to which they belong. Among the aims and objectives of these professional bodies are honesty, integrity, fairness and equanimity in all circumstances. Employers would probably agree that they hire their staff as much on the basis of the perceived character of the individual as on his qualifications, and so the letters after our names are as important in establishing that integrity as they are for establishing our capabilities.

These bodies have to maintain high standards, so that others can rely upon their integrity and the integrity of each individual member, and so these bodies employ strict disciplinary procedures. These disciplinary procedures are there to ensure that a standard of behaviour is maintained by every practising member of their organisation. They have the power to punish those who do not comply, or who consistently refuse to obey the regulations, by suspending or excluding them from the organisation, by removing from them the right to call themselves by their previous title or status, or by preventing them from using their former designatory letters. In some industries, this can be a powerful tool indeed, perhaps even more powerful than any punishment implemented by the courts and the judiciary.

The final aspect to be examined here is the role of religion, or religious morals and ethics. Such ethics can cause people to choose to work in certain fields, because their chosen trade or profession fits in well with their morals or the obligations they feel towards others. Thus there is a natural filtering into these professions of those people who are readily able to accept and observe their rules and conditions.

The same is often true of our peer groups. We tend, quite naturally, to mix with the kind of people we like, and with whom we can relate, and so we may filter into the peer group that most suits us rather than allow another group to change our values or behaviour.

So it is with religion. We will almost inevitably seek to attend a church whose beliefs concur with our own. Having accepted those beliefs, we must then follow all of them and not just those that we consider necessary or desirable. This kind of belief can impact on our ability to change or alter our value system, over a period of time or perhaps even instantly.

3.7   Other factors

There are some other factors that should be taken into account when considering our internal value system. We can override it in certain circumstances without suffering the same consequences as when we knowingly breach it. For instance, our value system tells us that we will not kill another person, yet perhaps a situation arises that causes us to kill someone in order to defend ourselves. We can override our value system in this case, because there is greater danger in complying with it than breaking it. Clearly this is an extreme example, but it does illustrate how a situation might alter our priorities.

We should consider whether changes in our society or in our personal situation really do alter our value system, or whether these things simply change our attitude as to whether or not we should comply with our value system. For example, people who have children of their own will probably regard issues such as child abuse or child education differently from those who are childless and who may have less defined feelings about such matters. Perhaps such a person felt less fervently about these things when they were childless themselves.

Have the circumstances altered our value system, or are we simply more acutely aware of the issues because we are more directly affected?

Changes in society might also affect either our value system or our attitude to it. We may feel quite strongly that we should not exceed the motorway maximum speed limit of 70 miles per hour. If every other driver felt similarly, the speed limit would seldom be exceeded, so everyone would feel comfortable about complying with that particular law. If, however, we perceive that most motorists flout the maximum speed limit, travelling instead at, say, 80 miles per hour, how would our own value system handle that? Would we drive at that speed, too, on the grounds that everyone else does it? Would we consider that law irrelevant, and ignore our own feeling that we should obey that law regardless of what others may do? This is a form of peer pressure, and is ultimately a question of whether we consider the speed limit to be important or trivial.

Our internal values are, for the most part, held subconsciously. Whilst they affect our conscious mind and we are able to consider them and whether or not we should exceed them, some of them are almost automatic. Inside each of us there is a 'robot', which controls our automatic movements, such as walking, breathing, eating and so forth, without the need for conscious thought. This robot can also have some influence on the way we think or react in certain circumstances. If we believe or have experienced something, the robot will develop our reactions in accordance with what we have learned. When we first begin to have driving lessons, we have to consciously think about each action and concentrate on how to steer or when to change gears. After a while the robot takes over, and driving becomes second nature. This is true of most things. The more often we perform a certain action, the more automatic it will become, because the robot can eventually take over and allow us to concentrate on something else.

If we decide to take any of these 'automatic' decisions away from our robot, then they will require more considered decision-making processes from our conscious mind.

Our value system will also be handled to some extent by our robot, because a considerable number of value-related decisions have already been made. For example, say someone offers you a cigarette, but you do not smoke. Your reaction, 'No thank you,' is automatic. It requires no thought on a conscious level, because the decision not to smoke was made previously and is merely being repeated. There is no need to go through the thought process, the choice process or the conflict process – there is no internal conflict, and the robot answers for you. If, therefore, your internal value system is well defined and your robot is well schooled in observing and obeying your internal value system, then you will find that you fall into fewer internal conflicts than those people whose internal value systems are less well-defined, or whose robots are not accustomed to obeying the value system automatically.

Some people have a very simple outlook on life, simple beliefs and simple understanding. Such people will tend to have a more defined robotic influence, whereas analysts, free thinkers and academics are trained to continually challenge theories, ideas and principles that are put before them. These free thinkers will have the greatest problems with internal conflicts, because they will need to consider every decision of any consequence very carefully, working through each option separately and applying each one to their internal value system. Each decision they make will have to be tested via a conflictive process within themselves.

In summary, then, we need to know ourselves well and we need to observe our internal value system. If we do, our internal conflicts will be reduced to the kind of 'approach–avoid' situations mentioned earlier. If we do not, then we may find ourselves facing constant moral and ethical dilemmas. By understanding and working with conflict within ourselves, we become more able to deal with conflict between ourselves and others.

4 Dishonesty and Self Deception

In the last chapter we saw how internal conflict can ravage an individual's self-worth and cause them distress. The evolving human psyche is not comfortable with this type of distress and so it offers defences to these internal conflicts. Whilst some individuals will learn to deceive themselves, or others, to protect themselves from such conflict, it may also be that their behaviour results from a more complex psychological process.

In addition to blatant dishonesty and self-deceit, there are two psychological processes that psychologists believe may be to blame for much of the conflict we encounter. These are:

- Cognitive Dissonance; and,
- Confirmative Bias.

Whilst studying for my QS diploma in the 1970s, the college insisted that all students had to gain at least one credit from an unrelated syllabus. I chose psychology and thus established a lifelong interest in human behaviour. When I first started researching this book, I realised that I needed further assistance, because I would have to consider the deep psychological and anthropological issues present in construction conflicts. Naturally I turned to those people I have known, or met, since my student days and who are practitioners in psychology and anthropology. I believe their help and research will help us all understand conflict a little better.

4.1 Plain dishonesty

This takes many forms, from the subtle to the brutal, and can be found in a surprising number of construction conflicts. Without making excuses for the less honest among us, we all have to admit that we face challenges with total honesty. Imagine you are at home and your wife has coloured her hair and it is an atrocious

Conflicts in Construction: Avoiding, Managing, Resolving, Second Edition. Jeffery Whitfield.
© 2012 John Wiley & Sons, Ltd. Published 2012 by John Wiley & Sons, Ltd.

(to you) shade that makes her skin look pallid. How brutal are you prepared to be? 'I'm sorry, dear, but that is quite shocking. You need to re-colour as soon as possible,' is one alternative. 'I don't know, I guess I could get used to it, but you looked much younger/prettier with your hair the way it was,' is another. One is more honest, but the other is more likely to maintain harmony in your relationship. Having said this, honesty is the best policy. Being open and frank may cause us immediate problems, but it can prevent major problems developing later.

A little while ago, I was the expert witness on a major international airport project in the Mediterranean region, and I was investigating the Contractor's delay. In forensically examining the records, I came across the monthly report for the period three months before Practical Completion was expected. The Contractor's report cited no problems and forecast that the project would complete on time. The project was partially handed over 9 months late and finally completed, in its entirety, 18 months late. The Contractor clearly knew that completion was not possible as reported, but he simply could not bring himself to report the truth. A long-running arbitration followed, which might have been avoided if the parties had worked together knowing the truth.

We could put that experience down to extreme optimism, if it did not happen regularly. Just a year or so later, I was working with the parties on a mixed use Development in the centre of Manchester, when the Contractor admitted that he was 33 weeks behind and not the 9 weeks being reported.

This type of avoidance lie is unacceptable and it will cause conflict, but it is understandable at a human level. What is less understandable is where conflict arises because of blatant dishonesty used to obtain a financial benefit.

As the Commercial Manager for an M&E contractor in the 1980s, one of my projects was a huge development in Hertfordshire. The project was overrunning and we could not make progress. Imagine our surprise when the main contractor blamed us for the entire delay and set off huge sums to pay for site establishment and security. The calculation was that the Contractor would bear half and we would bear half of the overrun costs. I argued against this, but to no avail. Then I changed tack. I contacted all 38 of the other sub-contractors on the site and found that they too had experienced set-off to their own accounts, ranging between 10% and 30% of the overrun costs. Representing the body of sub-contractors, I wrote to the Contractor's company secretary and informed him that if we did not have this money repaid immediately we would have to contact the police to report a significant fraud. The MD called me back very quickly and assured me he had no knowledge of the fraud. The Contractor subsequently removed the QSs and the PM.

Conflict is inevitable as soon as this type of conduct is discovered. Trust is lost and some have even been imprisoned after having been convicted under the Fraud Act. Whilst this is not a law book, let me briefly discuss here one important piece of legislation, as an example of the levels of honesty expected of us.

When I was studying for my law degree, the types of fraud perpetuated in construction claims were covered by the Theft Act 1968, where the slightly obscurely worded phrase *obtaining pecuniary advantage by deception* was included at S16.

Since the Fraud Act 2006 came into force, the offences we may see occurring in construction conflicts are more clearly described. Consider these words:

A person is guilty of fraud if he… Dishonestly makes a false representation, and intends, by making the representation – to make a gain for himself or another, or to cause loss to another or to expose another to a risk of a loss.

Does that sound like it might describe a claim you have seen in the past? Or indeed, one on your desk right now? It certainly describes many of the claims I have seen. The wording is clarified further, leaving little room for argument:

A representation is false if – It is untrue or misleading, and the person making it knows that it is, *or might be*, untrue or misleading.

That a construction professional can be guilty of perpetuating a fraud if he submits a claim for money/time (or to avoid LADs) without proper evidence is worrying enough, but to know that he is also liable if the statement proves to be untrue and he knew that it *might be* misleading or untrue, may be alarming to some.

Dishonesty in construction is like dishonesty in life; it will be uncovered and when it is, the consequences can be devastating. A couple of years ago I was acting as the quantum expert witness for a well-known retailer who was defending a contractor's claim. They were disputing the claim because during the project the sum in dispute was half a million pounds, whereas afterwards the loss was quoted as almost £1.5 million. Immediately suspicious about the claim creep, I examined the Contractor's records and discovered wide-ranging dishonesty. We found:

- claims for personnel who were not on the project;
- claims for plant not delivered to, or used, on site;
- sub-contract claims fabricated, allegedly under pressure from the Contractor.

I was instructed to prepare two reports; one for arbitration, and one for the police. The Contractor's MD was invited to a meeting and was faced with the police report first. The Contractor withdrew from the arbitration rather than face prosecution, giving up all claims, including any legitimate ones. If the claims had been honestly prepared, the Contractor would have received several hundred thousand pounds, but instead he received nothing.

Blatant dishonesty can create conflicts where none really exists or it can inflame conflict to the point where resolution becomes much more difficult.

4.2   Self delusion

When instructed by clients to assist in resolving a dispute, I am often puzzled by the attitude of the opposing party who seem to be in a state of denial about the *obvious* facts. At first glance, it might appear that the opposition is either naive or is lying. But this is not usually the case; often they are convinced their claim is sound. So, the question is how do people come to believe something that is provably wrong?

During my years of observing people involved in disputes, it has become clear to me that human beings do not cope well with failure, and so we subconsciously rationalise failure and rebrand it as success. Let me use a simple example that we can all relate to; the decision to buy a house. As construction professionals we are wily, we study the market carefully to ensure we choose wisely but, inevitably, the market does not offer us the ideal house at the ideal price and so we have to compromise. As we have noted previously, life often requires us to choose between two less than ideal options. This leaves us conflicted because we are constantly asking the questions; would I have found my ideal house if I had waited? Have I acted too quickly?

Add to this the unsettling fact that the house chosen seems to have defects or disadvantages (that other choices may not have had) and we begin to feel uncomfortable with our decision, and we may even suffer from 'buyer's remorse'. Even self-aware, well-educated and highly-qualified people cope poorly with the thought that they may have made a costly mistake. It can be difficult for individuals with a certain self-image to live with the possibility that they may have acted imperfectly. In short, they find it hard to forgive themselves for perceived errors. At the other end of the scale, there are some people who practice self-deception so widely that they can readily excuse themselves from any error. The remainder of this chapter is designed to help us understand why those who are, and who should be, self-aware often still fail to admit their own errors and cause conflict as a result.

For most human beings, dwelling on their errors creates unhappiness and dysfunction. To manage this situation, the human brain has developed automatic defensive mechanisms, which enable us to overcome these feelings of despair and regain full functionality.

4.3   Cognitive Dissonance

The theory of Cognitive Dissonance states that:

> contradicting cognitions serve as a driving force that compels the mind to acquire or invent new thoughts or beliefs, or to modify existing beliefs, so as to reduce the amount of dissonance (conflict) between cognitions.

The definition is a little academic, but it is clear enough with an example to help us. In the early days of the last century, a sect grew up around a prophecy that aliens would destroy the world in the mid-1950s. When the world was not destroyed as prophesied, the cult members were conflicted. They still believed in their *prophet*, but his prophecy was wrong. They could not live with this conflict for long and so, when a different *prophetess* spoke out in 1957 explaining that the aliens had spared the planet for their sakes, they accepted this reasoning with enthusiasm. In short, they now had a reason to go on believing.

You are all thinking, they were batty, and you may well be right; but this behaviour (cognitive dissonance) exists, in a less extreme form, in construction conflicts. A claimant will allege a cause of delay and will champion that cause of delay with vigour, without thoroughly investigating the true cause. Then, when the facts are researched and it appears that the alleged cause was not, as a matter of fact, the actual cause of delay, the claimant becomes conflicted. The claimant needs to believe that his allegation is correct, but the facts suggest otherwise. Luckily a consultant happens along and is able to produce a report that supports the claimant's original allegations, thus removing the conflict. The claimant can now happily believe he is right once again.

4.4   Confirmative Bias

A similar process occurs with the more common human behaviour of Confirmative Bias. The term Confirmative Bias refers to an aspect of human behaviour that induces us to reinforce our existing beliefs, often at the expense of the totality of the facts. To use the example of the house buyer mentioned earlier, this might work as follows.

We move into the house we have chosen, and find that the council tax band is lower than we initially thought, thus saving us £120 per year. This confirms our wise decision to buy this particular house and so we congratulate ourselves. We later discover that the house is situated in a post code that has a higher risk of burglary and our insurance premium rises by £240 per year. We attribute this overall loss to greedy, grasping insurers, thus validating our view that it is nothing to do with our otherwise wise choice.

Whilst as human beings we instinctively seek evidence to support our existing beliefs, we must try to ensure that we do not fall into that trap when supervising important projects. Whether we are pursuing or defending a right, we must examine *all* available data before forming an objective opinion. If we do not, a conflict is almost certain.

In construction disputes, we occasionally encounter Confirmative Bias and it tends to follow this pattern:

a)   One party makes an assertion as to why, in his opinion, the roof was late completing.
b)   His advisors seek confirmation that the roof was late for the stated reason, confining their research to the documents supporting the initial assertion.
c)   Usually this research proves the point and so it reinforces the advisor's own belief in the case.

Confirmative Bias is a natural human behaviour that is usually calming and harmless, but it ceases to be harmless when the behaviour migrates into the business arena where dispute costs are incurred because of poorly researched decisions.

In some extreme cases, if Confirmative Bias is to offer any real comfort, a party will have to actively avoid coming into contact with the truth. This is the adult equivalent of a child placing its hands over its ears and chanting 'La La La', so that it can avoid hearing bad news. In a recent radio interview by a former Presidential Press Secretary, it was noted that the US Presidential Press Spokesman is often excluded from important decision-making meetings so that he cannot inadvertently disclose the true facts under pressure from press questioning. This is a more cynical adaptation of the concept and the Americans have called it 'plausible deniability'. Unfortunately, it also appears in construction conflicts.

To follow the example at a) to c) above, our consultant is now convinced that his client's case is supportable, but:

- In two-party meetings, the Consultant becomes aware of facts that support an alternative *theory* as to why the roof was late, but he chooses to ignore it because it emanates from an opposition source, which could be perceived as being biased.
- The Consultant is instructed not to pursue research into the alternative theory (or chooses not to do so), as there is a possibility that he may undermine his client's case (and his own earlier advice) by uncovering evidence that the client's assertion was misdirected.

This approach of using selective evidence is commonly adopted by claimants and even by expert witnesses. In short, like the US Presidential Press Spokesman cited above, they hope to avoid the accusation of deliberate deceit by circumnavigating the truth and sustaining a comfortable ignorance.

In a court hearing a year or so ago, the opposing expert witness had the opportunity to examine his own client's documentation but failed to do so. As a result he was able to assert before the court, with a clear conscience, that a global claim was necessary because his client had said detailed records did not exist and he had seen nothing to contradict his client's view. Unfortunately for him, I had reviewed all of the documents and was able to explain to the Judge that the detailed documentation was in fact available.

So, what are we to glean from our limited understanding of these psychological processes? In summary, I believe we learn that:

- Human beings are programmed to defend their own psyche by the instinctive use of Cognitive Dissonance and Confirmative Bias.
- Some people will take these instinctive behaviours a step further and deliberately fail to recognise contradictory evidence.
- We are not inexorably bound by natural or instinctive human behaviours. Once understood, these behaviours can be overcome. By consciously seeking all of the facts and objectively examining them, we are able to arrive at the real truth and reduce conflict.

Armed with this knowledge, we may give opponents the benefit of the doubt when they have missed obvious evidence that undermines their case, but any such benefit of doubt will quickly disappear if they refuse to accept the evidence when it is plainly presented.

We have now reached the stage in this book where we have gained an understanding of conflict generally and our part in perpetuating or controlling that conflict. The next issue to consider is the role of others and whether or not our relationships with others can exacerbate conflict.

5 Interpersonal Relationships

There are a number of different reasons why we, as human beings, have the urge to conflict with others. Most of these reasons can be categorised as falling under one or more of the following headings:

- Anthropological factors
- Sociological factors
- Physiological factors
- Psychological factors
- Prejudice
- Personality types
- Urge to conflict.

Without delving too deeply into any of these issues individually, we do need to have a basic understanding of each subject so that we understand, and can cope with, conflicting individuals.

In discussing this chapter with academics from respected universities, it quickly became clear that conflict is a widely topical issue, on which a great deal of research has been carried out. However, much of this research merely acknowledges that, in certain circumstances and under certain conditions, people do conflict. In the past decade, there have been numerous attempts to explain why this is so, but most of this research relates to conflict between countries, rather than between individuals.

To identify not only the causes of conflict but how those causes relate to the seven categories listed above, it is necessary for us to extrapolate relevant information from some of the research already carried out, and apply it to individuals and companies working in the construction industries. We should also seek to use our empirical knowledge and published academic research into the construction industry, to understand the likely causes of conflict between people. Some of these causes will be easily categorised, and others will not.

Conflicts in Construction: Avoiding, Managing, Resolving, Second Edition. Jeffery Whitfield.
© 2012 John Wiley & Sons, Ltd. Published 2012 by John Wiley & Sons, Ltd.

I encourage you to take some time to study and understand these issues, as they are of paramount importance in gaining a full understanding of construction conflict and also of construction personnel.

5.1 Anthropological factors

There are many theories as to why human beings are so conflictive. One prevalent theory has us believe that Man is naturally competitive. It is thought that, in primitive or prehistoric societies, men were forced to compete with each other for even the basics of life, such as nutrition, sexual partnership and power. Those who were stronger were able to obtain for themselves the more satisfying lifestyles. The following example has been cited by anthropologists to justify this theory.

In primate societies, such as chimpanzee communities, the males are really outsiders. The adult females form a sisterhood, in which they care for one another and look after their young. Those of you who are of masculine gender would no doubt like to believe that this arrangement enables the male chimpanzees to defend the females and the young, and to forage for food. The reality, however, is far less romantic. When predators appear on the scene, the males will often run away, leaving the female group to defend themselves. The females seek out the food and also defend the young. In these circumstances, the males have very little to offer, except to provide the females with offspring.

A group of several female chimps will often have only one or two males who hover on the perimeter of their group. The females have no sense of loyalty to one particular male, and so they will procreate with the male who is either available or the most impressive in the masculinity stakes. The males, whose sole aim is to mate with as many females as possible, must therefore compete for their attention with courtship rituals and preening. According to how well the male performs, the females choose their temporary mate and the mate will fulfil his role.

Human socialisation is, of course, quite different, but theorists believe that many of these instincts are still prevalent in males and females in our modern society. Theorists cite the desire of males to build up their physical bodies at the gym as one example. We could equally consider the fear of baldness, from which some men suffer, as a fear not of losing hair but of losing their attractiveness to females, thus bringing an end to their potential procreative role.

The theory expounds that although some women will base their choice of men on their physical, economic and intellectual status, there is within that woman a subconscious desire to choose a man who is not only able to contribute his genetic offering, to provide healthy offspring, but to choose a man who is also able to support a family. Again, if this is true, modern men may find themselves having to compete financially for the woman they desire.

Such theories naturally lead to the belief that men are fundamentally competitive, and they could explain why some men like to be seen as 'macho' or as 'alpha males'. This pocket-book anthropology does, however, leave a number of questions unanswered. From our point of view, the most important of these questions is: Does the construction industry attract a larger than average proportion of the alpha males or macho, ultra competitive males and, if so, why? There are theories that seek to explain this too.

5.1.1 Social filtering

Anthropological studies, carried out in the UK and in the USA, have found that particular personality types are filtered into particular professions. There are a number of understandable reasons why this should be so, for example:

- My family tree has been researched back to the early 1700s and it indicates that generations often followed parents into factories, coal mines, dock-working and the professions.
- In the past, people were less mobile and often had to find work in the locality where they lived. This led to generations of farm workers in rural areas and generations of tin miners in Cornwall.
- Many people prefer to work with others who share their background, principles and objectives. This can lead to a peer group all entering a given trade or profession.
- In times of unemployment, often a job seeker's choice is limited and to survive they have to accept the jobs offered rather than their preferred occupation.
- Education and educationalists channel students into professions that seem to suit the students' abilities and personalities.

However, these reasons alone cannot explain the high correlation of certain groups in a given industry. Studies carried out on dockworkers in Wales found that many of the workers had certain basic common characteristics. The same can be said to be largely true of construction workers too. A study of construction workers in the 1970s in the USA was published in the book *Royal Blue – The Culture of Construction Workers* by Herbert A. Applebaum. In the study, the author found a high level of agreement between construction workers on issues such as job satisfaction, use of humour at work, productivity and job security. Some other social issues, such as heavy drinking, swearing and the treatment of women were also discussed and again large areas of agreement were identified. The study suggests that a certain type of individual would enjoy the varied but uncertain life of a construction worker, and that such individuals seem to be attracted to it in great numbers.

This gives rise to the belief that social filtering must occur within the construction industry. If, therefore, we can accept that a certain type of person would prefer to be a quantity surveyor and a different personality type might prefer to be an architect, is it too big a leap to imagine that people with a naturally aggressive or conflictive nature might be filtered into a management role in a conflictive industry?

Of course, generalisations, whilst capturing similarities, tend to obscure differences. Everyone is different, and categorising people into a single group will inevitably cause us to overlook some non-conforming aspects of their behaviour, but as humans we do need to categorise things in order to deal with them. The construction industries are often seen as being macho, aggressive and conflictive and so anthropological academics suggest that many of the people attracted into the industry will show these characteristics in a greater proportion than a similar sample of, say, accountants would show.

5.2   Sociological factors

The aggressive/competitive factor is only one of the elements contributing to the urge to conflict with your fellow man. Sociological factors also have to be considered. These include ethnicity, race, religion, class, age, culture and many other group issues, and each of these is mingled with individual personality and predilection.

If we were to study these in depth, we would need to refer to learned books on the specific subject matters, but we can highlight some of the more important social influences.

5.2.1   *Experiential influences*

Our attitude towards dealing with the other people will be crafted from our experiences and how people have dealt with us in similar circumstances. Human beings learn many things through their experience of life, both positive and negative. We learn that pain, physical or emotional, is undesirable and so we instinctively avoid the possible causes of such pain. Physically, we keep our distance from dangerous items such as excessive heat, poisons and sharp objects; emotionally, we avoid relationships and avoid allowing others to know too much about us. Thus, we may feel safer, but this kind of emotional withdrawal can cause us to encounter difficulties in relating to other people.

An abused child will try to work out what behaviour causes his carer to hurt him and then avoid that behaviour. If that does not work, he will try to avoid his carer altogether, recoiling from the problem, not facing it. Often, however, the cycle does not end at this point. Someone who has been abused will often retaliate against society, targeting others less powerful than themselves. Imagine the scenario where the father returns from work and shouts at his wife, she shouts at the child and the child kicks the dog. It is this cycle of abuse that causes bullying and racial abuse. People of low self-esteem will sometimes conclude that they can raise their self-esteem by exercising control or power over weaker members of their community or unpopular social groups.

5.2.2   *Interpersonal influences*

Our relationships with others are largely based upon what we see in our early years at home. The example set by our parents is important. If we fail to develop good interpersonal skills at an early age, then we may develop protective instincts that we believe will cushion us from failure. The reasoning is self-defeating but understandable and works in the following way. Mr A has poor social skills and finds conversation difficult. When engaged in discussion with Mr B, he answers only with 'yes' and 'no.' Mr B assumes that Mr A is not interested in conversation and avoids him in future. Mr A feels rejected and his self-confidence wanes. After a period of time, he will actively avoid conversing with others, further reducing his social skills.

Such people will reflect this attitude in one or more of a number of ways:

- They will begin to despise themselves.
- They will begin to despise the other person.
- They will begin to despise everyone.

We need to have the ability to recognise this type of behaviour as being symptomatic of a person whose interpersonal skills are very weak or perhaps even non-existent. We should also be prepared to help these individuals to develop good social skills. Failure to do this may ultimately result in our having a difficult person to deal with, perhaps even a sociopath. If, on the other hand, we take the time to help that person, we may find that we gain a friend for life and benefit the whole of society.

The alternative to this lack of confidence can be seen in the audition stages of programmes such as *The X Factor*, *American Idol* and similar talent shows. Often the contestants singularly lack the proclaimed talent and yet somehow manage to be the only person who has failed to notice. Occasionally a family group will express confusion or outrage at the failure of their offspring to qualify for the next round, whilst we, as objective outsiders (and voyeurs), wonder how these individuals have become so deluded that they will argue with three professional judges that they were singing in tune. The key often lies with parents, friends and teachers who have lied, kindly, about the quality of their listening experience, when the contestant is vocalising to music and is straying only occasionally, and inadvertently, in the direction of the tune.

It is plain to see that both a lack of confidence and a surfeit of confidence can soon result in conflict.

5.3   Physiological factors

Our instinctive and acquired behaviours are not the only factors that govern our ability to deal effectively with others. Physical well-being has a part to play too.

5.3.1   Physical attributes

Appearance, physique and persona all influence the way in which other people react to us. Few people would wish to argue with a 15-stone muscle-clad body-builder, yet they would have fewer qualms about trying to impose their will on a frail old man.

5.3.2   Health

When we are enjoying good health, we may feel able to discuss the pros and cons of an issue for many hours with a colleague. If, however, we are suffering from a common, but painful, ailment, such as a migraine or an upset stomach, our tolerance level will probably be greatly diminished. Our normally easy-going temperament may vanish along with our sound stomach and be replaced with a vitriol to match the acidity swirling around in our solar plexus.

Hormonal imbalances can also affect our reasoning, patience and temper; to the degree that such chemical imbalances have been cited as mitigation for out-of-character behaviour in court cases around the world. A similar negative impact on behaviour may also be prevalent in someone experiencing alcohol abuse or drug abuse issues.

5.3.3   Sensitivity

Comments or remarks made to someone, even in good health, can be misread. As noted in the introduction, a comment made at a heated site meeting, such as 'Keep your rug on,' is usually understood as 'Calm down,' but to a man with a real or imagined hair loss problem, it may appear as a gibe.

The answer to most of the problems that we face on the physiological front is to gain a better and more complete understanding of the people we deal with and their physical challenges. Unless we know someone very well, we would be well advised to think carefully about how we use humour, repartee and collo-quialisms. This brings us to the final factor in causing unnecessary conflict between people – the Psychological Factor.

5.4   Psychological factors

5.4.1   Perception

Any abbreviated study of psychology will quickly turn out to be a study of perceptions or understandings. Psychologists conduct tests, during which they show ambiguous pictures to different individuals. Those individuals will often see different things in the same picture. Take the most common of perception tests below:

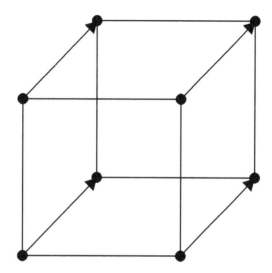

Is the box facing away from you or towards you? The arrows don't help at all, do they?

A perceived difference can cause conflict on a construction project, even when a proper analysis shows that there is no real difference between the parties at all. Let me share another anecdote (facts altered for anonymity) to help illuminate this point.

Around ten years ago, a Counsel for HM Treasury employed me as a quantum expert witness to help resolve a dispute on a major new building between a government department and the Contractor. The source of conflict was this; the Contractor believed he was due to an additional £1.2 million spread over 132 unagreed variations. I looked at the variations and found that they had been agreed in principle but not in value, but neither party seemed to appreciate this.

Furthermore, I discovered that the reason that the 132 variations were not agreed, as to value, was that the government's QS had asked for evidence of labour rates and had never received them, so he disallowed the VOs by placing them in the unagreed column. My research showed that, of the 132 variations, most comprised agreed invoiced materials and hired plant and less than 50% of the variation claim value was for labour. More alarming was the fact that the base labour rate was agreed in the Contract and so only the claim for a 20% on-cost was unproven. A quick calculation showed me that only £120 000.00 was really in dispute. Once the parties were informed that their dispute was so much smaller than perceived, they settled amicably.

You may find it hard to believe that disputes can become so entrenched when they could be avoided entirely if the facts were understood, but I can assure you that most cases settle when independent expert witnesses narrow the differences through analysis and research. To further illustrate my point:

About five years ago I was asked to act as quantum expert witness on an offshore dispute, something of a specialism for me, and I accepted the assignment gladly as the difference between the parties was £24 million. The difference between the parties was shown in the pleadings by way of a PDF version of a spreadsheet prepared by a barrister. Two weeks into the case, I could only find a maximum of £16 million in dispute. In a tetchy meeting with lawyers, all of whom thought I was a bit daft, I tried to explain that the difference between the parties was not as significant as it looked, because £8 million of the claim was actually pleaded in the alternative. The Claimant had pleaded both a day-work case and a prime cost case for the same work. On the electronic version of the spreadsheet summarising the Defence, both cases were included in the totals. My client had, apparently, always been pre-pared to pay around half of the sum remaining in dispute. So, by the time I pointed out to the opposing expert that there was also significant duplication of alleged losses, the parties' expectations were soon close enough to allow them to settle amicably.

Perceived differences are not always real differences, and cases that could be negotiated away often run to formal proceedings, because the real difference between the parties is exaggerated.

5.4.2   Women and men

Anyone who has a close relationship with a member of the opposite sex will have found that men and women can see things differently. In the last 20 years, many books have been written about the difference in psychological makeup between women and men, especially in terms of communication. Learned authors have sought to explain how these differences can cause avoidable con-flict, specifically books such as *Men are from Mars and Women are from Venus*. In one of the first of these books; *That's not what I meant – Men and Women in Conversation 2* – Virago Press, 1992, the author, Dr Deborah Tannen, explains that a great deal of the conflict between men and women arises from their different perceptions and viewpoints. Dr Tannen says:

> Relationships are sometimes threatened by psychological problems, true failures of love and caring, genuine selfishness – and real effects of political and economic inequity. But there are also innumerable situations in which groundless allegations of these failings are made, simply because partners are expressing their thoughts and feelings, and their assumptions about how to communicate, in different ways. If we can sort out differences based on conversational style, we will be in a better position to confront real con-flicts of interest – and to find a shared language in which to negotiate them.

With women and men working together more than ever in the construction industry, we need to concern ourselves with proper well thought out commu-nication, which crosses the gender divide.

5.5 Prejudice

Just as we should try to understand the challenges that people face physically, we should also be careful not to become a challenge to someone from a different social grouping. If we are working with people from other cultures, religions, ethnicities – whether in their country or our own – we can make great steps to avoid conflict by understanding their beliefs or lifestyles.

An ability to accept people as they are, without considering them to be of increased or decreased worth, is a valuable asset in avoiding conflict. In over 35 years in this industry, I have worked with people from all of the major religions, people of every ethnicity and colour, individuals who had different physical and mental abilities to my own and people whose sexual orientation differed from mine. The way to avoid the spectre of judgement or prejudice is to concentrate on your common goal and your similarities, whilst not being blind to your differences.

5.6 Personality types

Differences in understanding are not just an issue between different genders, race and ethnicities, but also between differing personalities.

As we have already mentioned, conflicts can arise from conflict of interests or they can arise from within. The greatest challenge presented by any conflict is that which relates to individuals and to the different personality types with which we work. This section is dedicated to identifying the different personality types and their reactions in given circumstances.

5.6.1 People

No matter how large or small the company you work with, or for, happens to be, its most important resource will be the people who work within it. People are creative, emotional, have experience and are also the reason for any construction process being undertaken. Those people are also unique individuals, as we discussed earlier. Individuals will act differently when faced with the same set of circumstances. Our reactions will be coloured by our past experience, our state of mind at the time, our well-being and health, our point of view – and even what type of day we have had.

Some or all of these factors can have a significant effect on the outcome of any negotiation, discussions or conflicts in which we may be involved. These factors do not simply apply to us, but they also influence the people with whom we will be dealing and those with whom we are in daily contact. Some of these personal factors will be beyond our ability to control and therefore we can make no impact upon those influences, but we can still find effective ways of managing difficult people, if we think analytically and if we understand the personality groups involved. In order to recognise the different personality groups, we need to review the characteristics of each group and see how those characteristics relate to the individual with whom we are dealing.

5.6.2 Stress

One thing that all personality groups have in common is that they are often subjected to stress. There is now a much greater recognition of stress as a real problem within individuals. Some employers, however, still refuse to acknowledge that stress is anything other than a sign of an individual's weakness. An unsympathetic response in these circumstances would be to tell the sufferer to snap out of it, fully expecting it to be done. We should recognise that for some, this is a wholly unrealistic expectation. A far more useful approach would be to have available a stress management programme, to enable company employees, at all levels, to recognise stress and to manage stress when they identify it.

A certain amount of stress is not only necessary but inevitable in our daily lives; however, should stress spiral out of control, it will affect how we perform at home, at play and at work. One of the first symptoms of an overstressed individual is their inability to concentrate. We may notice also that an individual is making errors of judgement, that they suffer a lack of motivation, have a reduced efficiency, perhaps their perception changes, and they may lose their creative or innovative input into the organisation.

Stress affects people in as many different ways as there are different personalities. One person under stress might develop a headache, perhaps even a serious migraine; another might simply become ill-tempered or snappy, while yet another might give in to pressure when subjected to it from others. We need to recognise when these signs are present in our colleagues, in ourselves and in our opponents when we are in a conflict or a negotiation. We should all try to take positive steps forward in alleviating stress before it affects the sufferer and creates a conflict where none is necessary. The sooner the signs of stress are spotted, the easier it is to deal with that stress.

Recognition of stress as an illness is necessary, so that we do not underestimate the effect that it can have on an individual's performance. Many employers implement stress management programmes when they want their employees to perform at their very best. It could be disastrous in a negotiation or conflict situation if the team involved simply do not perform properly because one or more of them is over-stressed. Basically when a person is under stress they will tend to make irrational, quick decisions without reasoning out the possibilities as they normally would. This could be devastating to the outcome of an expensive dispute in the construction industry.

5.6.3 *Personalities*

In order to avoid unnecessary conflict with our counterparts, we need not only to be able to control ourselves but to know something about them too:

- What is his personality type?
- What is your personality type?

The answer to these questions will be difficult to discern, unless we first understand how the characteristics of the different personality groups are displayed. So, to answer these questions, we first need to identify the different personality groups into which people can be broadly classified. Whilst every individual is unique, as we have already explained, there are certain behaviours that tie us in with one or more of the basic personality types. In this section, we will examine the extremes of each type.

	People Oriented	Task Oriented	
Assertive	Entertainer	**Ruler**	**Aggressive**
Passive	*Admirer*	Analytic	**Submissive**

The first direct comparison that can be made on the subject of personality profiling is that of assertiveness. There are those people who are assertive, and those who are non-assertive or passive. Even within these broad definitions, there are yet more variables. For example, there is a difference between passivity and submissiveness, and there is a difference between assertiveness and aggression, these usually being extremes of the same definition. Secondly, we need to establish whether we are a task-oriented person, or a people-oriented person. Finally, are we a leader or are we a follower?

We should now take a look at each of these variables, to see if we can recognise each personality type within ourselves and within others. If we examine the personality model, we will see that there are four quadrants and within those quadrants there are four personality types. The four differing personality types are: on the task-oriented side, the analytic and the ruler, whereas on the people-oriented side we find the admirer and the entertainer. If we examine the quadrant in a different way, then the people above the assertiveness line, whether assertive or aggressive, are the entertainer and the ruler, whereas the people found below the assertiveness line are the analytic and the admirer. As we said earlier, you may well be a mixture of more than one of these basic personality types, but you should be able to use this model to identify which of these quadrants you more readily appear within.

Knowing your own personality type will also help you deal more effectively with those people that you have identified as being in other quadrants. Each quadrant and each personality type has both good and bad points. We should try to eliminate as many of the bad points as possible within our personality group, whilst emphasising the good points. We should now look at each individual quadrant to identify the characteristics that influence those individuals, so that we can identify ourselves and others and categorise them accordingly.

- *Ruler*: A ruler is a person who is largely task-oriented. They always appear above the assertiveness line and may well be on the aggressive side of assertive. The ruler will generally want to be in control, and will be able to make on-the-spot decisions. Often efficient and well-organised, they will tend to be good time managers. The ruler does, however, hate time wasters and does not suffer fools gladly. Rulers also like to get straight to the point.

  As a result, there is also a tendency for the ruler to be pushy and blunt, almost to the point of being insulting, particularly when under stress. The ruler may also be intimidating, and will often be seen by others as a know-it-all. Rulers may be impatient or intolerant of those who do not share their views or see things from their viewpoint. The ruler may also try to make their point or show resentment of others by sniping at them, especially in a group environment. The greatest fear of a ruler is that they are no longer in control, and this fear will bring out all of the negative characteristics in their personality.

- *Analytic*: The analytic loves precision and accuracy. He or she will collect as much information as possible before making any decision. Weighing each option very carefully indeed, the analytic will veer away from the approximate. If you ask them the time, then they will tell you exactly, often to the second. The analytic will be painstaking, studious and a little slow, because they will need time to examine every aspect of the problem before making a decision. Analytics will strive to be accurate and precise with everything. As we know, there are very few occasions within the construction industry where there is a finite single answer and where all information is readily available, and so often an analytic will find himself incapable of making a decision.

  On the negative side, the analytic will also tend to be a complainer, largely because they hate inefficiency and inaccuracy. They may also be described as negative, a wet blanket, finding problems with every solution proffered. The analytic may say: 'That will never work because...' and if they remain unchecked, that might become one of their stock phrases.

Analytics may also be afraid that they will be blamed if something goes wrong. The analytic's greatest fear is being proved wrong. This is why they require so much information before making a decision.

- *Admirer*: The admirer is a warm, empathetic person, with skill in dealing with others. Because of this tendency to relate well with others, the admirer is often referred to as a relater. They put relationships above programmes and even projects, and their priorities are getting along with others, showing sensitivity and consideration. The admirer is able to see another's point of view more easily, because they really do seek to understand, to see what others see and, effectively, to feel what others feel.

  Admirers will often hide the truth, because they do not want to risk hurting another's feelings or losing a friend. Under stress, the admirer may appear indecisive and even defensive. The admirer will try to avoid confrontation, and will sometimes bend the rules or overlook important details. In a negotiation, the admirer might give in too easily rather than cause a conflict. The admirer wants to be valued, to feel needed and important, and to make others happy. Their greatest fear is confrontation, or having others turn against them. The admirer is on the passive side of the equation and their counterpart on the assertive side of the quadrant is the entertainer.

- *Entertainer*: This personality type is also prevalent in the construction industry and has a people priority tendency, but the entertainer is also on the assertive or aggressive side of the quadrant. Whilst entertainers like to be admired, to be the centre of attention, and want others to look up to them, they will not sacrifice everything for that end. Nonetheless, the entertainer wants to be at the centre of everything. Acting spontaneously, making decisions quickly and often on impulse, the entertainer is enthusiastic about everything, and wants you to be enthusiastic too. They are open about their feelings and they may exaggerate to get your attention. The most important thing to remember about the entertainer is that they need to be needed.

  On the negative side, the entertainer may go to extremes in order to be noticed, seeking prestige and recognition at all times. The entertainer tries to do too many things at once and never stands still. Concerned with making a good impression, no matter what, the entertainer's greatest fear is to be put down by others they admire or like, or even worse, to be completely ignored.

As we have said already, you and others will fall into one or more of these categories. Understanding the people and their characteristics is important in order to avoid conflict, but why should conflicts arise merely because people have different characteristics? It may already be obvious to you how these personalities can impact on one another, thus causing conflict. If not, you can examine the most likely clashes in a later chapter.

5.7 The urge to conflict

When all of these factors have been considered, it is perhaps more surprising that there is any harmony at all in the construction industry rather than the fact that conflict is prevalent. We cannot expect to remove the urge to conflict entirely, but we can identify areas where we conflict inadvertently or unnecessarily and eliminate those problems first.

In order to assess how these very general principles can be found on a construction project, we analyse a typical construction contract in Chapter 6.

6 Anatomy of a Construction Project

In the foregoing chapters, we have examined the component parts of a conflict. We now know that there are many variables that may contribute to contention between the parties on a construction project. Identified in our reading so far have been three main potential areas of conflict, namely:

1. A conflict arising from differing interests or positions;
2. An internal conflict arising from a breach of values;
3. A conflict of personalities or of individuals.

Each area has been considered in a simple way, but we now have to examine them in the context of a complex construction project. In reality, do any or all of these issues arise in a typical construction project? This chapter should answer that question definitively.

6.1 Overview of the project and the parties

For the purposes of this chapter, I have chosen to base the analysis on a hybrid of major projects, which progressed to international arbitration. All cases used have been duly settled by the Tribunal, Mediation or by two-party settlement talks. Whilst those involved in these projects may recognise some of the incidents, the account below is a fictionalised version, which protects the anonymity of the project and those involved.

There are many reasons for choosing these particular projects, but chief among these are realism and variety. The projects involve real-life examples of civil engineering, building, M&E, process pipework and other services.

6.1.1 The client

The client, a European Energy Company, wished to construct a new gas fired power station in the Middle East, to match supply more closely to

Conflicts in Construction: Avoiding, Managing, Resolving, Second Edition. Jeffery Whitfield.
© 2012 John Wiley & Sons, Ltd. Published 2012 by John Wiley & Sons, Ltd.

demand in the region. The client believed that it was essential to get their electricity and desalinated water to market quickly and so the client, whom we shall call Middle East Generating (MEG), decided upon a fast track, turnkey project in the United Arab Emirates. Their aim was to obtain, as quickly as possible, an efficient and economical combined cycle power plant that used the exhaust gases from the gas turbines to produce steam for a further steam turbine.

6.1.2   *The feed contractor/design contractor*

Whilst MEG's own employed engineers could produce an outline design quite ably, they were not able to put together a complete working design for the plant.

The designers chosen by MEG were experienced and expert in the power generation field, and were designer/contractors in their own right. Design Engineering Services (DES) – as we shall refer to them – were to develop the outline design into a fully serviceable design, by producing detailed 'Approved For Construction' drawings. The Invitation to Tender (ITT) would invite the Contractor to further develop detailed drawings and workshop drawings from the AFC drawings provided. The design covered the whole project, which was to be let including:

- gas Turbines
- HRSGs (boilers)
- pipework
- other mechanical and electrical services
- controls and instrumentation
- buildings
- roads and drains.

DES could not complete the design in the time available before ITT, but as this was a fast track project, MEG agreed that the later mechanical design stages could continue whilst the early civil engineering works were underway on site. The Client and Designer were of the view that the main plant, although complex, was in fact quite compact and so the works should present little scope for major conflict.

Between them, MEG and DES did foresee one problem; as the greatest labour value was in the mechanical engineering scope of work, it was essential that the company carrying out that work was a top tier contractor. So, it was decided by MEG and DES jointly that they would invite tenders from internationally renowned contractors with pipework manufacturing and installation experience gained in the power and nuclear environments. It was also felt that only household names would be capable of achieving the client's brief in the time allowed

6.1.3 *The Contractor*

Although at ITT the outline design would be complete, the detailed design would have to be developed further and so the contract would be for 'design development, detailed drawings, procurement, supply, commissioning and setting to work' the power plant. Unfortunately, relatively few well-known companies were interested in tendering for the scheme, and it was eventually awarded to an internationally renowned power engineering contractor who would develop the design and also project manage the work. The bulk of the site work would be packaged and sub-contracted to civil engineering contractors, steelwork contractors, mechanical and electrical sub-contractors, and instrumentation/controls specialists.

The successful contractor, we shall call International Powerconstruct LLC (Powercon), was to employ directly all necessary contractors.

Powercon employed the following sub-contractors:

- Mid-east Civil Contractors Ltd (MCC)
- Pipework and Engineering Contractors (PEC)
- Steel and Concrete Construction Co (Steelco)
- UAE Mechanical and Electrical Co (M&ECo)
- Cosgrove Controls Ltd (Cosgrove).

6.2 The pre-contract period

The client, MEG, were in the business of generating electricity in many countries and providing wholesale electricity to the Middle East, UK, Germany and France. As such, their designers were familiar with the design of plants and their operability, but were less familiar with buildability. Almost immediately, DES, the designers, discovered that their brief from MEG was incomplete. DES consistently asked for details of the major plant packages (turbines, boilers, etc.), but MEG believed they had given DES enough information for the ITT and refused to give more details. DES accepted the situation and, without full knowledge of the plant, designed the civils, the builders' work and the balance of plant (connecting pipework, cables, etc.), hoping to compensate for the incomplete brief provided by the client.

Very quickly the design fell behind programme. The client blamed the designers for the delay and the designers counter-claimed that the delay was due to the reluctance of the client to divulge information on the major plant itself. One example cited by DES was the turbine supplier; until it was known who was supplying the turbine and what its size, weight and footprint would be, DES were reluctant to finalise the design of the foundations and slab.

The fluctuations in the price of gas were causing concern at MEG's regional headquarters in Dubai and their Finance Director insisted that the initial capital be committed in this financial year or the project funding would slip two to three years. Afraid of losing the financing, the MEG PM called a meeting with the designers to avoid the postponement of the plant build.

The MEG PM demanded that the design be sent out to tender within the original timescale. DES pointed out that the design was far from complete and the Contractor may make huge claims for time and money if the work scope was uncertain when he tendered. After some prolonged discussion, it was decided to proceed with the fast track philosophy and send out the incomplete design on a target price for the work tendered and a schedule of rates for the undeveloped scope. Having agreed on this method of proceeding, the MEG PM contacted his Finance Director. The Finance Director, an accountant by profession, said that MEG would only proceed on a fixed price contract. If that was not possible, then the scheme should be mothballed.

The MEG PM now had a dilemma; should he proceed on a fixed price basis against all professional advice? If he did, his job was secure for the duration of the contract, but if it went wrong, what then? On the other hand, if he did not proceed, would he still have a job? An internal conflict loomed.

The PM decided to go ahead and tender on a fixed price lump sum contract and, in order to protect himself against it all going awry, he decided to amend the contract conditions to increase the Contractor's risk, whilst at the same time diminishing his own.

Acting on internal legal advice from Europe, MEG decided to procure the plant, not on any of the standard forms of Contract, but on a heavily amended and less than appropriate hybrid form.

DES, who compiled the tender package, warned that this method of procurement would create problems later. MEG's PM instructed the designers to send out the package as it was, without drawing the Contractor's attention to the level of incompleteness of the design. The designers sent out the package as instructed. A number of contractors received the tender enquiry and soon the client was inundated with technical queries. The client passed the queries to the designers, who answered them as fully as they could.

The designers sent a fee note for this service. The client's PM refused to pay the excess fees and insisted that the technical queries were an inherent part of the agreed design fee. DES, the designers, denied this. They had spent the three month tender period answering queries that arose because the drawings that were sent out were incomplete, and contractors quoting a lump sum price wanted as much data as possible.

The contractors all sought extensions of time for the submission of their tenders, but none were granted. Powercon priced the work keenly, as they needed to increase their turnover in a tight market. However, their price was not the lowest and it exceeded the budget by 58%. MEG were concerned about the tender levels as every tender exceeded budget dramatically, one by over 100%. Worse than this, perhaps, was the fact that every tenderer had heavily qualified his price. After prolonged negotiations with the lowest tenderer, the parties failed to reach agreement. The client called in Powercon and said that if they removed their qualifications and agreed to complete the design within their price, the project was theirs. Powercon saw the opportunity for securing over £500 million worth of turnover and the onerous contract was signed.

6.2.1 The analysis so far

There is nothing unusual in this set of circumstances. Similar events happen every day in the construction arena. So, have the seeds for a conflict already been sown? Let us examine the problems arising to date.

- *Time*: The project is being rushed. Mistakes will be made and those errors will be more time-consuming to resolve later. For example, if the 'one size fits all' foundations do not suit the turbine supplier's bespoke needs, thousands of cubic metres of reinforced concrete have to be removed and recast. Furthermore, the Contractor has not fully comprehended the work scope, but has offered to achieve a completion date that may not be achievable.
- *Honesty*: MEG's PM is deliberately withholding information, which the Contractor needs in order to fully identify risk and price the work properly, because he needs to reduce the initial price if the project is to go ahead.
- *Fairness*: The client has altered the contract terms to increase the Contractor's risk and the Contractor, not knowing of the issues in design, is not in a good position to quantify that risk. In addition, the client has deliberately maintained a tight tender period, which will prevent the Contractor from investigating the scope of the problems too deeply.
- *Money*: The Contractor has concerned himself too much with the turnover he will achieve and has not considered sufficiently the costs he may incur by eliminating his qualifications. So, are the seeds sown? I think so, but let us look further to see if they have settled on fertile ground.

6.3 The contract period

As the design commenced, the Contractor, Powercon, became concerned at the amount of design that had been left for them to complete. Some areas of the work, principally controls and trace heating, were hardly designed at all and had to be researched, developed and then designed. Powercon employed specialist sub-contractors to carry out these works, but at a cost.

The contracts with the sub-contractors were intended to be 'back-to-back' with the main contract, so that all obligations of the Contractor became obligations of the sub-contractors too. This simply did not happen. The sub-contractors each refused to take a share of the risk taken by Powercon on what was an obviously incomplete design.

The fast track proviso fell by the wayside, as Powercon refused to begin constructing until a design freeze was imposed. Their reasons were sensible. They argued that until the major plant was procured, DES could not finalise the foundation drawings, nor could they design the cable and pipework routes without the final plant design and, as all cables and pipework would be on racks with significant foundations, even the excavation could not commence.

MEG ordered Powercon to begin the enabling works such as roads, utility buildings and service supply trenches. Powercon did as they were instructed. For the next nine months, holes and trenches littered the site, preventing access for the major plant, which had to be craned in over completed ancillary buildings at huge cost. Work was done piecemeal and there was no continuity.

Powercon also encountered 'buildability' problems immediately with both the controls and the trace heating. Under the sub-contract conditions, Powercon were obliged to pay the sub-contractors for the research, development and design of workable systems. When Powercon sought to recover the excess design cost from MEG, the client, MEG purported that it was all included in the tendered work scope under the general heading 'Design Development'.

Whilst they would not admit to the fact, MEG secretly agreed with Powercon that the design was flawed and was significantly further from completion than had been stated in the tender enquiry. This much became clear later, because MEG used the evidence of the flawed and incomplete design in their own defence to the DES Notice of Arbitration. But, rather than concede this point honestly, MEG instructed their quantity surveyors to refuse to allow additional payments for the 'increased' design scope.

Powercon knew that if their sub-contractors found out about the non-payment from MEG, they would cease work, so Powercon remained silent on the issue.

Contract progress was slowing because every contractor was spending time waiting for the completion of the inadequate design before construction could commence. MCC were late with the foundations, Steelco fell further behind with the building work and, more importantly, PEC still had insufficient data from MEG/DES to complete the rack and pipework design.

In an effort to accelerate progress, Powercon decided to instruct PEC to commence fixing the pipework to a point close to where the plant would be, before the whole pipework design was fully completed. Naturally, when the plant arrived, there were significant changes in the pipework design, which made some of the installed pipework redundant. Powercon confronted the client and accused them of providing misleading and inaccurate outline design information. The client pointed to Powercon's responsibility for 'design development', and said that must include rectifying errors too.

By this time the project, supposedly of 11 months' duration, was 8 months behind schedule. Powercon had spent more on design development than had been allowed and had received less than expected in cash flow owing to the slow progress. Powercon were also receiving significant claims from sub-contractors.

The Contractor asked the client for a lump sum payment to cover their excess costs, and the client refused. A meeting was arranged and the MEG Finance Director (who spoke only French) attended, as did Powercon (who spoke only German and English). No money was forthcoming from MEG, only threats, and so Powercon left the site to consolidate the design. They said that when the design was totally complete, they would return and complete the works. The client, who was secretly pleased at this action because the price of their gas supply had risen dramatically, making the plant temporarily less viable, concurred, and the site was abandoned.

For the next nine months, design costs escalated in sub-contract offices around the world, and within the Powercon design office, without any significant cash coming in. The conflicts that had arisen so far were about to be elevated to new proportions, as people became frustrated and angry.

6.3.1   The analysis so far

Again, these circumstances are far from unusual in the industry. Conflicts will arise, but it is the manner in which they are addressed that decides the success or failure of the project. Let us examine the Contractor's problems.

- *Money*: It became clear almost immediately to Powercon that they had taken on more than they had allowed for in their tender. But, rather than address the problem there and then, they agreed to defer a solution while they, Powercon, spent money and the client sat back waiting for them to solve MEG's problems. Powercon eventually suspended the works, but only after eight months' loss of productivity and eight months' lost costs. A bold decision made too late?
- *Honesty*: When problems arose, Powercon chose not to disclose them to the sub-contractors. This resulted in rumours circulating around the site and it created a lack of trust within the sub-contractor organisations. Questions arose, the key question being: 'Why are Powercon not paying us?' The sub-contractor's prejudicial conclusion was probably predictable: 'Perhaps they are keeping the money for themselves.' It is a fact of life that an ignorance of the real facts results in an increase in suspicion and mistrust, then the void left by the absence of fact is filled with rumour. All of this provides fertile ground for people, so minded, to engage in conflict.
- *Conflict of Values*: The MEG PM would have served both parties better by setting a realistic budget and timescale for the works, but he feared his project would be cancelled and he would lose his job. So, internally conflicted, he acted in his own favour.
- *Win-Win Outcomes*: MEG did not need the plant imminently and so could have agreed with the Contractor a revised programme and new budget that would have helped both parties. The opportunity was there for a new beginning. But this would have meant the PM facing criticism/discipline from his Finance Director and so it did not happen.

6.4   The dispute period

When the design was eventually completed, Powercon wrote to the client asking for a variation to cover the extra design work and they also requested an extension of time. After protracted negotiations, and under a new MEG PM, work on-site commenced again upon receipt of a lump sum payment for unagreed extras.

The pipework re-commenced and, until it was complete, the trace heating could not commence. Until the trace heating was complete, the insulation could not be completed.

The mechanical and electrical contractor had now carried the bulk of the cost of his own redesign and variations. He was now distinctly unhappy about the lack of payment. Mr Carr, their accountant, wrote to Powercon saying that, unless they were paid, they would cease work and sue for their outstanding account. Powercon cited the sub-contract terms, which did not allow such actions, but Mr Carr alleged that the Contract had never been signed.

A review of the documents showed that a formal contract had never been concluded. It was also clear that some important qualifications in the M&ECo tender bid were still outstanding at the time of suspension. Mr Moore of Powercon (an Entertainer) refused to speak to Mr Carr (a Ruler), because he perceived him to be a 'troublemaker'. Mr Carr simply wrote to Powercon's Managing Director (also a Ruler), who in turn instructed Mr Moore to open discussions and resolve matters amicably.

Mr Moore arranged the meetings and wrote a memo with the words: 'Get me all the dirt you can on these people; I'll teach them to go over my head.' Inevitably the meeting broke down and the M&ECo contractor left site. Powercon did not tell the client.

Progress slipped yet again, and the project was now over 18 months behind.

Powercon called in a US-based claims consultant and appointed a new PM. This new team adopted a very tough approach and in the short term achieved some unstable progress. All sub-contractors were back on site and a meeting was arranged with the Finance Director of MEG, the client.

MEG was unprepared for the onslaught that faced them, and for the mountain of evidence amassed against them. The Finance Director agreed a completion deal, which gave Powercon more money and more time if they achieved a new agreed end date. The former MEG PM was severely reprimanded.

The next day, the MEG quantity surveyor was summoned to a meeting with his long-standing friend, the new MEG PM, and was asked to undervalue the Powercon work in the monthly valuations. He explained that he was unable to do so, as his professional ethics forbade it. The PM suggested he find a way to slow payments down or they would view his lack of co-operation as a sign that his company no longer wished to work for MEG. His decision was pre-empted by MEG's decision to call Powercon's On Demand Performance Bond, citing slow progress.

The loss of banking facilities, worth over £50 million, led to the failure of Powercon, who slipped into receivership without warning.

The project ground to a close some 21 months late, massively over budget and with disputes raging from all sides. The M&ECo contractors are suing

Powercon (in Receivership), claiming that there was no contract, and so they are demanding payment on the basis of *quantum meruit*, which they perceive to be the same as cost plus.

Powercon's Receivers have issued Arbitration Proceedings against MEG and, using the evidence provided by sub-contractors, have forced MEG to make a significant sealed offer, which Powercon have refused. MEG are unaware that Powercon have been in discussions with DES, who have offered to give evidence on the fact that MEG knew the design was hopelessly incomplete. MEG are counter-claiming for the huge costs of finishing off the project with new companies. As you might imagine, the case would take years to resolve.

6.5   Summary and analysis of the dispute

The reader will note that the facts cited above avoid the mention of the normal, everyday conflicts over measurement, variations and late payment, which continually beset construction projects and of which there were many. To allow clear analysis, the expert witnesses on both sides concentrated on the major issues and found that on this project:

- MEG failed to be honest in the procurement process, choosing to defer conflict or hoping that some circumstance might arise to save them.
- Powercon failed to understand or properly price the work before signing the contract.
- MEG failed to be fair in the settlement of legitimate delay and cost issues.
- Both parties concealed problems and allowed conflicts to grow, rather than manage them as they arose.
- All parties permitted coercion, emotional involvement and personal attacks to divert their attention from the real issues.
- Everyone failed to grasp the opportunity to turn a disadvantage into an advantage, for example, co-operation on design and buildability.
- No-one recognised that a suspension was in everyone's interest and could have been used very productively
- Under stress, the parties acted dishonestly and tried to impose unethical behaviours on principled professionals.
- Both parties refused to be realistic in their expectations, as to what to expect from the troubled project.
- All parties refused to communicate when that was the only possible way to resolve the issues.
- The costs to complete the works were unnecessarily inflated by pulling the bond and sending the Contractor into receivership.

This chapter is intended to show the ways in which conflict, in all of its forms, can arise in a large construction project. All of these, and more, happened on one particular project and so we need to be aware of the catastrophic effect our behaviour can have on the project and the industry. The Contractor that went into receivership in the real case had been trading for almost a hundred years.

Ways to avoid these conflicts are discussed in the next chapter, with advice on how to reduce and manage conflict following in later chapters.

7 Twelve Steps for Reducing Conflict

The construction industry, as we have seen, can be a fertile ground for conflict. There are many different reasons for this, but I hope that it is clear that, where possible, the risk of any event deteriorating into an actual conflict or dispute should be reduced. There are a number of simple positive steps that can be taken when dealing with individuals or companies, to reduce the frequency of conflicts. These measures, when taken together, can greatly reduce the risk of your discussions becoming conflictive.

The purpose of this chapter is to bring all of these steps together in one place, thereby enabling you, the reader, to familiarise yourself with them quickly. You should be able to see results as soon as you begin using these steps.

7.1 Step 1: Communicate with precision

Most of the work we do takes place within groups and organisations, and we inevitably have to spend a great deal of our time communicating with others in order to achieve some kind of goal. Our success will depend on how well we are able to communicate. In general terms, we communicate on three levels:

1. Content (the spoken word)
2. Tone (the way we say the words)
3. Visual (body language).

We will deal in more detail with these separate aspects in the next chapter. In this chapter, we will concentrate on the first of these, the actual words that we use to express our ideas to others.

Structured communication can be a factor in avoiding conflict. One of the main causes of conflict is misunderstanding owing to a lack of clear communication. When explaining your ideas or setting out your position to others, be sure that you also state your objectives clearly and concisely. It usually helps to have your thoughts written down beforehand. You might also consider preparing a copy of these objectives to give to your counterpart, so that he can see instantly and clearly what you are saying. Be sensitive and avoid giving the impression that you are making demands or trying to pressurise your counterpart. You do need to be firm in stating your case, but you must also be polite and to the point. Above all, it is wise to show that you are willing to be flexible and that you are open-minded, approachable and that you are prepared to listen to what they have to say.

Avoid ambiguity, and speak plainly; they can take bad news. Always state your point of view clearly. You must be prepared to speak candidly and honestly, even if it feels uncomfortable to do so. Too often individuals try to conceal their real intent by using ambiguous words or by deliberately concentrating on minor matters to deflect the other party from the important issues.

A form of contract issued by a nationally renowned contractor was deliberately worded with imprecise words and meanings to provide an argument or defence should they fail in their obligations. Over many years in the industry, spent considering whether or not to recommend the acceptance of contract documents, the most valuable question I have ever asked is:

What does this clause really mean?

I then incorporate my answer into the Contract or Agreement. A contract, any contract, is simply an evidenced agreement. To agree to something, you must understand it. Without agreement, real agreement, there can be no contract.

In Chapter 6, the client chose to hide his real meaning in a reworded contract. He chose not to communicate clearly, because he believed that he would benefit from the ambiguity of the wording. Only dispute specialists benefit from ambiguous wording!

Some conflicts can be avoided if everyone involved understands the bargain that they are striking. By communicating our requirements precisely, we avoid that most common of conflicts in construction, the one that arises when someone says:

…but I read it this way, your interpretation is incorrect.

Even the humble brick needs to be properly described to avoid misunderstanding. To describe a brick adequately, we need to specify the following:

- size and shape of the brick;
- special features of the brick;
- colour or finish of the brick;
- whether the colour or finish is on more than one side;
- whether the colour or finish is surface only or throughout;
- who makes the brick and their contact details;
- the name or reference number in a manufacturer's data.

If you then explain the reason why this brick has been specified, you give the other party an opportunity to communicate back to you his experience of that brick and its suitability for your purpose.

Communicating with precision in the written and spoken word is that single most powerful tool in avoiding a conflict of ideas or beliefs.

7.2   Step 2: Listen and consider attentively

Communication is, in essence, a two-way process, so in addition to presenting your own case, you must be willing to listen to what the other side has to say. Too often, we hear what others are saying without really listening. We may be trying to compose a suitable response, yet how can we respond adequately unless we understand all of what they have said? Others are not truly communicating, because they are simply waiting for a pause in the speaker's narrative to enable them to interrupt and have their say.

Effective listening is a skill, which must be cultivated. Really listen to what the other person is saying, and try to see things from his point of view. When he has finished, express your understanding of what he has said.

Resist the temptation to interrupt whilst someone else is speaking, even if you believe them to be wrong or if they are complaining about you or your company. Take notes, let them have their say first and only then offer your response. When you do respond, try to avoid being negative.

The other side deserve to have their views heard just as you do, so make sure you allow them to do so. Conducting two monologues will achieve nothing except conflict, so effective listening is necessary on both sides.

We should also read attentively, looking carefully at what is written. Identify not only the contents of the document but its tone also. Does the document appear to be honest and open or does it give the impression of hiding something? Use discernment in examining documents sent to you by others and in your own written and spoken communication. Try to read between the lines and gather the real message.

In Chapter 6, the Contractor saw the contract documents and knew that they had been amended, but he was not 'listening' attentively to the message. The Contractor needed the job and so failed to pick up the sub-text of the documents. In this case, the Contractor had the warning signs flashing before his eyes but ignored them and instead examined the revised wording at face value. He should have asked himself why the contract wording was being altered in the first place, by using this checklist:

- Has the wording been changed for increased clarity?
- Has the wording been changed for increased ambiguity?
- Has the wording been changed to reduce your risk?
- Has the wording been changed to increase your risk?

Employing solicitors to alter standard contract wording costs thousands, if not tens of thousands of pounds, and so clearly the contract presenter must perceive that there is a value in changing the wording. That purpose will almost inevitably be to their benefit.

Later, in Chapter 6, the Contractor wrote to the client asking for additional time and extra money and the client simply refused the request. The client failed to 'listen' to the tone of the Contractor's message in the letter, whilst the letter contained the words:

> We request an extension of time and reimbursement of our additional expense…

It did not say, but meant:

> Look, we both have a problem here. Your building is going to be late and will cost more if we do not talk. Why don't we meet and agree how much more time and money is proper under the circumstances.

The reply, which answered the written words, did not address the unwritten message in the letter. Clearly, the Contractor should have followed step 1 and communicated precisely, but where that does not happen, an attentive client can still avoid a conflict by listening to and considering the real words left unspoken.

By developing an attentive attitude to both oral and written communication, we can identify and resolve problems before they become conflicts.

7.3   Step 3: Think before speaking

Many of us respond automatically to what we see or hear, without really considering the message our response sends to others. We may even have decided how we will respond before we have heard what the other side has to say. This will obviously waste time, as they are sure to repeat themselves until they are sure you have listened and understood. Do not say anything without thinking carefully about what you really wish to convey, and without considering first what your counterpart has said. You may even require a few minutes of thought before you respond. If you do, take the time and explain that their observation requires some deep consideration.

Do not be afraid of silence. It is extremely valuable, because it allows time for both sides to digest what bas been said and it provides valuable seconds in which you can formulate a more thoughtful response. It also shows respect for the other side when you really concentrate on what they have said. It is far better to spend a few moments formulating a suitable reply than to say something that you might regret later.

Choose the words you use with care. You should consider using key words in your conversation, words which are most likely to motivate the other person or trigger within him positive emotions. The most powerful words you can use are 'you', 'yours' and the person's first name. Consider which of these examples is most likely to motivate, or gain the most favourable response.

> I'm very impressed by your company's progress on site.
> Ron, you handle this project so well. You're a real credit to your company.

There are, in addition, key words that have been proven to have a positive impact on any conversation. Research has found that the following are the 20 most persuasive words in the English language:

easy	save	results
effortless	keep	benefits
simple	retain	discovery
new	secure	guarantee
love	protected	promise
money	health	free
safe	strength	

These 'power words', used carefully, can stimulate positive feelings in the other person and are more likely to lead to a favourable result. Avoid using negative words; especially 'no', 'can't' and 'won't', and make liberal use of the other person's name. By thinking before you speak, and choosing your words with care, you will increase rapport and greatly reduce the risk of conflict between yourselves.

Another thing to consider when responding is the tone or timbre of your voice, and whether that can be moderated. Tony Blair and Margaret Thatcher both had voice coaches, despite being proficient public speakers in their own right.

We used to have a dog at home, and if I shouted at him harshly and said: 'You are a wonderful dog, I love you,' he would cower in the corner. Alternatively, if I said smilingly: 'Come here, I am going to beat you with a stick,' he would run to me and lick my hand. His response was based entirely on the tone of my voice, rather than the actual words spoken.

7.3.1   Tone is so very important in responding to others

By considering our words carefully before speaking, we avoid creating unnecessary conflict. When we are frustrated or angry, we often respond more harshly than we would otherwise.

The sentence:

> I think that this should be rewritten to see whether you can improve the summary.

may become:

> Not again! For goodness' sake, take it away and try to get it right.

No-one wants to disable a subordinate by undermining his self-confidence. We want the job done correctly by a subordinate who learns from his errors.

A wise old man once said to me, 'Be careful what words you use because, once spread, they are as hard to unspread as butter on bread.'

Careful consideration of our tone and our choice of words combined with control of our emotions will help to avoid conflicts and stop those conflicts that are inevitable from escalating.

7.4   Step 4: Take time to build relationships

This step should follow on naturally from the previous one. Let the other person see that you are sincerely interested in him and his cause. There are specific ways in which you can do this on a subconscious level, such as pacing, which we will discuss in detail in the next chapter. There are also some more obvious ways, such as using the person's name frequently. Also, by showing a genuine interest in others, the chances are that they will show an interest in you, too. Each of us has a basic need to be liked, and we respond more favourably to those who show that they like us.

Establishing a rapport and building a relationship reduces the likelihood of serious conflict, because most people prefer not to contend with friends or with those in whom they have invested time and emotion. You will, however, need to be aware of the other person's personality type when attempting to build your relationship with them.

If the person is an analytic, for example, he might view social 'small-talk' as a waste of time, or he may be suspicious of your motives in wanting to know things about him. A relator, on the other hand, may seize such an opportunity as a chance to talk about himself, pushing aside the real reasons for your meeting. The relator is also the most intuitive of people and therefore the most likely personality type to be able to spot when you are being insincere, so be sure to say only what you really mean.

Another benefit arising from building a good relationship with your counterpart is that he will be more likely to want to deal with you or your company in the future. Time spent in getting to know each other a little better is never wasted, but will inevitably bring positive results for both sides.

Often overlooked, the opportunity to mix socially and find out what people are like away from work is crucial to a relationship. No matter how enthusiastic we are about work, we are only there a small proportion of our 168-hour week. Understanding other people by building a relationship outside the working environment helps us to ignore the uncharacteristic bad behaviour and look beyond it to see the real person underneath. By knowing people more intimately, we will not conflict as often, and minor conflicts will be resolved as they are overshadowed by the commitment to an on-going relationship.

7.5   Step 5: Be honest in your dealings with others

A major cause of conflict is distrust. If the other person believes you to be honest and fair, he is much more likely to trust what you say. It can take time, of course, to earn another's trust, but it will take only an instant to destroy it. Avoid telling lies or half-truths, or making ambiguous or misleading statements. We sometimes hold back the truth because we are afraid of being ridiculed, or of offending others. We generally wish to be accepted and liked by others, and so we naturally wish to avoid anything that could damage our popularity. Yet we can tell the truth in these situations, if we choose our words carefully. Being honest does not necessarily mean being blunt or brutally frank. We can tell the truth tactfully; letting the other side know that our intention is not to offend but to offer constructive help. A good relationship with the other person will make this much easier. Bear in mind that being less than honest now is more likely to cause offence in the long term.

Even the most skilled deceivers are caught out sometimes, and the resulting damage can be very difficult to repair. Trust, once betrayed, can be impossible to earn back. To let the other person know that you are being honest, maintain eye contact as much as possible. Do not stare – this can make others feel uncomfortable – but look them in the eye when you are stating your facts, and emphasise your sincerity by your tone of voice and by your posture.

It is important to avoid platitudes and clichés, and you should especially beware of using phrases such as:

To be completely honest with you…

This will trigger a subconscious reaction in the listener that says you are either about to lie or have been lying up to that point. There are a number of similar phrases, the most common of which are:

- 'Frankly speaking…' – They are not being frank at all.
- 'That's a good question…' – They do not know the answer.
- 'I've been meaning to call… '– They had forgotten all about you.
- 'I'll have to get back to you on that…' – They are completely stumped.
- 'I'll just look at your file…' – They cannot remember who you are.

Avoid using phrases such as these, and listen for them from your counterpart. They signal, in most cases, a movement away from total honesty.

In Chapter 6, we saw how the client refused to disclose all information and how this dishonesty led to a breakdown in the relationship between client and contractor. Lies are always found out sooner or later, especially contractual lies, and then your credibility is called into question.

Honesty in procurement is essential to the success of a project. All members of the team should be equal partners in having access to relevant information. Deceit today merely defers the conflict to some not-too-distant tomorrow.

By dealing with problems honestly, we will have to face up to some uncomfortable decisions, but these decisions can be made together as a team, and the result will be a joint approach that reduces the potential for a conflict to arise in the future.

7.6   Step 6: Do not dispute trivial matters

Stick to the main issues once you have begun a discussion. Being pedantic or making personal comments will not advance your cause or foster any kind of reasonable relationship with your counterpart. Your aim should be to have constructive discussions on the key issues, rather than trying to score points. Unfortunately, there are a few individuals for whom winning trivial arguments or demeaning their opponents is a great triumph. Do not allow these people to sidetrack you or drag you into pointless arguments over insignificant issues. The result will be conflict over minor points, whilst leaving the main issues unresolved.

Very often, there are trivial matters that need not be disputed, because they have little or no bearing on the outcome. In my first arbitration, I was an expert witness as to the value of a construction claim. The claim had some validity, but relationships had become so strained over a legal point that neither side wished to compromise. The two parties had disagreed over the wording of JCT 63 clause 23(1), Statutory Undertakers. The Contractor argued that GPO telephones were not a statutory undertaker if their works simply passed through the contract works, as they did. The client disagreed. The point was referred to a preliminary hearing and the Judge decided in the Contractor's favour. The cost of the parties, their solicitors, junior counsel and QCs came to many thousands of pounds. When I came to consider the value of the claim arising from this point, I found that there was none. No value, not a penny, rested on the issue. Where there is no need to conflict, do not. It merely fuels the fires of future conflict.

This advice relates also to negotiations. The less there is in dispute, the greater the momentum for settlement. After all, if 20 items are disputed, no-one wants to delay settlement on one single sticking point.

By avoiding unnecessary argument we will maintain goodwill and reduce the likelihood of further conflict considerably.

7.7   Step 7: Look for common ground

In any conflict there will, almost certainly, be some points on which both sides can agree. Make maximum use of any such agreement by listing the points of agreement and using them to show your counterpart that you are making progress and that you are both seeking the same result. You might also choose to emphasise that you are confident about reaching agreement on the remaining issues. If the other person believes that you agree on most matters, he will be more enthusiastic about reaching agreement on the remaining matters.

If there is very little common ground between you, there are other ways of achieving it. Pacing is specifically designed to build common ground, and is a powerful tool because it works on a subconscious level (see Chapter 9). Finding as much common ground as possible is more likely to encourage a feeling of co-operation. Minimise the differences between you as far as is possible, and use positive and encouraging phrases to lessen the risk of conflict.

Always allow the other party to a disagreement an honourable and face-saving way out. It may be superficial, but they will be happy to take it and will be grateful. In a very difficult final account negotiation, the client's quantity surveyor had said emphatically that the Contractor was entitled to a mark up of only 5%. This was clearly wrong and the evidence could have been presented there and then to defeat his emphatic remark. The Contractor's site agent, a wily old construction manager, recognised the difficulty that the QS would have in retracting his statement and so approached the matter differently. He said:

> You are a man of principle. If I were to prove to you that your 5%, although reasonable in normal circumstances, was inappropriate here, you would listen, wouldn't you?

The QS agreed to being a man of principle and now had an honourable way of conceding the point if necessary.

If we look for common ground fervently enough, we will find ample opportunity for joint progress. In a dispute over the development of redundant land, we have a prime example.

A well-known supermarket chain purchased a piece of derelict land on the edge of a town and applied for planning permission to build a large new store. The council had made a public statement in the local elections that the land would be used for social housing. A conflict loomed. The matter was referred to local mediation and an architect suggested that the supermarket could be built with social housing above, the social housing being partially funded by the supermarket. After discussions, it happened that a trust formed by the supermarket built the social housing with no cost to the local council.

Common ground is there if we look for it! By jointly looking for solutions, rather than arguing against each other's proposals, we begin to solve problems more readily. After all, there are twice the number of brains thinking about the problem, and the human mind is an inventive and powerful tool, so let us try to use the power of partnership to deflect conflict.

7.8   Step 8: Recognise and avoid prejudice

We cannot avoid differences. They are a fact of life and these differences can actually enrich our lives. Life would be boring if everyone was the same in every way.

In construction, as in other areas of our lives, we will come into contact with widely differing types of people, and the way in which we treat them says a lot about us.

There are now many more women in the industry, as well as people from different races and ethnic groups. We will also meet people from varying socio-economic groups and people with opposing political views. Patronising or mistreating anyone from a so-called 'minority group' will inevitably lead to conflict.

Everyone deserves to be treated with equal respect. If you, the reader, fall into what can be classed as a minority group, you will no doubt have experienced negative attitudes and prejudices many times. If this is the case, try to be tolerant of those who show such prejudice, but show them that you are indeed their equal.

Those who would patronise or look down on others need to recognise that, whatever cultural differences exist between individuals, these have no bearing whatsoever on whether or not a conflict can be dealt with successfully. Everyone deserves respect, as well as fair and equitable treatment, not only at work but in every other aspect of their lives.

Prejudice is not the issue it once was in the industry, but it does occur from time to time and it needs to be eradicated entirely. Prejudging others on grounds of race and gender will not only foster bad feeling, but will also cause you to underestimate your counterpart.

With all of the conflict inherent in the industry, we must avoid the most avoidable conflicts, those involving gender and race.

7.9   Step 9: Express your understanding

Conflicts often arise because others believe that we do not understand their point of view. Full understanding of what your opponent is saying is vital in any discussion. Always ensure that you have listened carefully to your counterpart, and that you are aware of everything he has said. You may also repeat back to him the facts as you have understood them. Backtracking in this way is not, as some people believe, a waste of time. It will actually save time in the long run, by decreasing the likelihood of misunderstandings.

It also confirms to the other person that you really were listening to what he had to say, and it reinforces the feeling that you respect his point of view and have taken it seriously.

If there are any areas of ambiguity, do not be afraid to dispel them by asking direct questions. Let the other person clarify any points about their case of which you may be unsure. Do not make assumptions or leap to conclusions of your own. Always try to defer making a decision or reaching a judgement until you are sure that you are in possession of all of the facts.

Make sure, too, that the other person completely understands what you have said. This is more likely if you have followed the advice given in Step 1 and have clearly communicated your point of view. Complete understanding reduces the risk of conflict.

In dealing with construction claims, a working knowledge of this step is invaluable. If you fail to acknowledge the other side's view, they will continue to repeat it until you do. This can cause fruitless and irritating conflict as both sides become increasingly fractious. One side thinks, 'How many times do I have to say this to make him understand?' The other side thinks, 'Why is he repeating this over and over? Does he think I'm stupid?'

When faced with a claim that needs a response, do not do that which is so common and counter immediately with your view. Reflect on their view first. Consider their points and show by reiteration and by asking questions that you want to fully understand their position. A good deal of time, effort and ultimately money, is wasted in proving points that have already been conceded by the other side in construction disputes.

Expressing understanding of another's point of view helps to reduce tension and conflict. At least you will both understand the differences between you, so there is no conflict over what is in dispute.

If we refuse to communicate, we cannot understand and so cannot resolve those differences that are easily resolved. Keeping the channels of communications open is always of importance in avoiding and reducing construction conflict.

7.10   Step 10: Control your emotions

We should all recognise the fact that there are some people in this world with whom we cannot, no matter how we try, build a close relationship. There are many reasons why this is so, but they are unimportant. In all cases, we should be friendly and polite, even when we must differ.

Many construction conflicts arise because those involved allow their personal feelings and emotions to override the issues. Some people have a tendency to dwell on past mistakes, or perhaps, there is a grievance – real or imagined – which they hold against us or our company, or vice versa. In meetings and negotiations, we need at all times to stick to relevant and current issues and not dwell on past experiences.

Emotions will inevitably cloud any issue, so we need to put aside any personal differences that may exist between ourselves and our counterpart. Never involve yourself in an emotion-driven argument over historical issues. Apart from the wasting of everyone's valuable time, it will serve only to force the other side more firmly into opposition against you.

Some individuals will not be openly abusive, but will snipe at you with sarcastic comments or innuendos. Do not fall into the trap of responding in like manner, as this will perpetuate the arguments. Defuse their anger with words that are likely to elicit sympathy, such as:

I'm sorry you feel like that.

In all cases, you should remain calm and stay on the right track. Once the other person has let off steam, he may well calm down, too. If you feel unable to remain calm yourself, ask for a five-minute break and use the time to gather your thoughts rationally. This will also allow the other side time to cool off.

Another way to defuse a potentially explosive confrontation is to allow the other person a way out. Sometimes a person will back himself into a corner. If this happens, and he is unable to see an easy way out, he may resort to abusive or violent behaviour. Always leave the other person a way out, or build him a bridge over which he can retreat gracefully. This will help to reduce tension, and lessen the likelihood of an embittered conflict. Remember, there is no place in a properly managed conflict for bitterness, abuse, revenge or winning at all costs.

We should try to remember that the bouquets in this industry are awarded for successful projects, not for successful arguments.

By carefully controlling emotions, we prevent a minor conflict over issues becoming a major conflict of personalities. Conflicts of issues are usually overcome more easily than a conflict of personalities. Hurt feelings can be inflamed throughout this project and carried into the next one.

Always be aware of purportedly cantankerous and hostile opponents. They are often very good actors. So, whilst controlling your own emotions, call attention to the emotional outbursts of others. Tell them in an assertive but controlled way that you are not prepared to be abused, verbally or otherwise, and leave if necessary, after assuring them of your co-operation when they calm down.

The control of our emotions is an important step in controlling and avoiding conflict.

A few years ago, I travelled to the USA for an expert witness meeting. Whilst I was in the air, our initial reports were exchanged. When I arrived at the opposing expert's offices, it was clear that I was not welcome. Eventually the opposing expert came in and, boiling with rage, threw my report at me. He was grossly insulted that I had criticised his work, in writing and in a document disclosed to the arbitral tribunal. The fact was that I was criticising his client's claim and not his report. I had not yet seen his report. I soon realised that they were probably one and the same. Nothing I tried would calm the situation and the result was that he picked up a chair and hurled it in my direction. The ensuing noise brought staff rushing in and he was ushered away. When the opposing expert had calmed down and we were able to discuss matters unemotionally, we found we had reached a high degree of agreement on principles and that we differed mostly on methods of valuation. Subsequently we agreed that this was a matter for the Tribunal. We are now good friends, but for a few minutes it seemed as if the conflict between experts would hamper the resolution of the conflict between the parties.

7.11   Step 11: Apologise gracefully if you are wrong

No matter how professional we may be, we all make mistakes from time to time. Every one of us is human, and therefore capable of making errors of judgement or simply misunderstanding others. We may like to believe that we are infallible or invincible, but no-one really is. No matter how hard we try, there will be times when we upset or offend another.

No-one likes to admit that they are in the wrong. It hurts our pride and may appear to damage our self-esteem. We can rapidly rebuild our diminished self-image if we learn to accept the mistake, or the fact that we have offended someone else, and apologise sincerely. A gruff 'sorry' muttered under the breath is not enough. A genuine and heartfelt apology is what is required.

Do not make excuses such as:

I didn't mean to do that, but I couldn't help it because…

The 'but' qualifies the apology and puts the blame on to something or someone else. Face up to the mistake, and admit it without reservation. Far from damaging our reputation, such an admission may even impress the other person and gain their respect. Someone who is honest and open about their mistakes, and sincere in the apology, is more likely to make friends than to lose them. If, in addition, you are able to make a restitution of some kind for the error, then you will have advanced your reputation still further.

In the bestselling book by Ken Blanchard on *One Minute Management*, he cites the one-minute apology as a powerful tool. His formula is roughly as follows:

Peter, I am sorry for the way I acted, I just want you to know that I do not always act like that. It was out of character and I will try not to act that way again.

The formula is: recognise the individual, say sorry, reinforce the fact that you are well intentioned and then commit to eliminating that behaviour in future. Only a very churlish opponent could refuse to forgive you after that and, if he did refuse, he would be in the wrong.

As a general rule, people are very forgiving and will have great affection for someone who apologises with sincerity. In court, a few years ago, I encountered a barrister of some fame in advocacy. An ageing quantity surveyor was in the witness box and the barrister was in full flight…

So, Mr Mitchell, you failed to get the valuation done on time and then you failed to release the retentions as the contract directs. What did you say when faced with that allegation at the meeting of the 12th January?

Well Sir, I said I was sorry. I was then and I am now. It was my fault and I accepted that without question. I naively expected to be forgiven, not to be sued.

The barrister sat down and all eyes fell upon the plaintiff, who looked at his shoes with some intensity.

Apologies defuse conflicts and disarm those in conflict. Use them sincerely and often to avoid contention and reduce disharmony.

7.12 Step 12: Accept apologies gracefully if others are in the wrong

As we have discussed, admitting and apologising for our errors takes courage. When the other person is in the wrong, it is important that we treat them with respect and accept their apologies generously.

We all know how hard it is to apologise. Do not rub salt into the wound by being impatient with those admitting their mistakes, by taking offence unnecessarily, or by saying 'I told you so.' Make it easy for them to apologise. Be friendly towards them. Show them that you have no hard feelings, that you forgive them and that you understand how they feel. Empathise with them because, after all, you know how it feels to make mistakes.

Build their self-confidence, and help them to put things right, if necessary. This will help to strengthen the relationship between you, reducing the risk of conflict over the error.

Remember that it does take courage to admit mistakes and apologise for them. Respect this courage. Do not make things more difficult for them. Be generous, and you will earn their trust and respect.

By accepting sincere apologies gracefully, we encourage good behaviour in the future. If we refuse to accept apologies, we will miss out on possible friendships and we may discourage those who have erred from apologising in the future.

Magnanimity in success is as important as good humour in defeat. Make sure that you always accept apologies generously and conflicts will generate less quickly.

By using these 12 steps, we should be able to reduce conflict within the industry but there are still those people whose urge to conflict is strong. These people will create a conflict where none was apparent. Therefore, learning to deal with these people is important, and we examine this aspect next.

8 Reducing Conflict

In the previous chapter we discussed ways of avoiding conflict in the construction industry. The application of those principles will also reduce the intensity of a conflict where such a conflict is unavoidable. This chapter is dedicated to the amelioration of interpersonal conflict, which can either create conflict where none exists or turn a minor conflict into a major battle.

If we are to set out to diffuse or reduce conflict, we will need to recognise that we always deal with conflict through the medium of human interaction. Regardless of what we do or want, we can only achieve our intended result by dealing with people. If we need ten cubic metres of ready-mixed concrete, then we order it through people, and it is delivered by people. Should the concrete be below specification, we do not berate the concrete, we berate the people who supplied it. People are always the activator in getting things done.

When conflict arises, we should remind ourselves that we do not deal with companies, we deal with individuals, individuals who have their own personal goals as well as corporate goals. The country's largest contracting companies are merely collections of individuals. When sub-contractors say that the Main Contractor is disputing their final account, they generally mean that an individual QS working for the Main Contractor is disputing the account.

8.1 People, people, people

A construction conflict is generally characterised by an individual or group in one organisation conflicting with an individual or group in another organisation. Knowing how to deal with people successfully must, therefore, be of paramount importance in dealing with conflicts and disputes. This facet of management, interpersonal relationships, is often undervalued in the construction-related industries. How we, in the industry, react towards one another as individuals will have a pronounced effect on the amount and severity of the conflicts that

Conflicts in Construction: Avoiding, Managing, Resolving, Second Edition. Jeffery Whitfield.
© 2012 John Wiley & Sons, Ltd. Published 2012 by John Wiley & Sons, Ltd.

we encounter. To assist in reducing the amount of unnecessary conflict caused by human interaction, we need to consider the following:

- interpersonal techniques
- individual perceptions and personalities
- tactical behaviour.

Understanding the principles established here and then implementing strategies to accommodate these principles will improve your relationships dramatically, so do try to adopt at least some of the suggestions.

8.2 Interpersonal techniques

We have already mentioned some of these techniques in passing, but let us now consider these powerful tools in more depth.

8.2.1 Smile

A warm smile communicates to others that we care about them. It may even be contagious. Philosophers over the centuries have proclaimed that a smile has properties beyond a simple warm welcome. Physiologically, a smile can change the way we feel and therefore the way we respond. Call centres instruct their staff to smile on the telephone, because they believe that their customers can tell when someone is smiling over the telephone line. Basically, a smile can change the way we and others feel, reducing tension and the urge to conflict. If you are still unconvinced, remember that frowning uses more muscles than smiling. So smile and save energy.

8.2.2 Use names often

Using someone's name communicates two separate messages; firstly, that you care enough to remember, and secondly, that the person is important to you. The researchers tell us that no sound in the English language is quite as sweet to us as the sound of our own first name. Try using someone's first name, where it is appropriate to do so, and watch for the results. It is highly likely that they will use your first name, too, building a rapport. The closer we get to people, the less chance there is for misunderstanding and conflict.

8.2.3 First impressions

As the saying goes, you only ever make a first impression once, so make it an impressive one. People judge quickly on a whole variety of often irrelevant criteria; for example, dress, manner, accent, gender, appearance and behaviour. Ensure that you perform at your best when meeting someone for the first time. First impressions are often difficult to overcome.

8.2.4   *Show interest in others*

If you can only master one interpersonal technique, let it be this one. This is by far the most powerful tool you have for influencing your own life and the lives of others. People love to talk about themselves, usually, and they love people who listen while they do so. Listen attentively to others. Watch out for hidden messages in the words or tone. Ask questions to show that you have been attentive, and listen with genuine interest to the answers. Share yourself and your experiences with others, too, after they have had their say, and you will be surprised at how quickly friends can be made. Importantly for us, friends seldom conflict as harshly as enemies.

By building relationships with others in the industry, we can not only avoid conflicts but benefit from future work, and obtain greater enjoyment from our employment.

8.3   Perceptions and personalities

8.3.1   *Perception and reality*

By now life will have shown you that some types of people are naturally more conflictive than others. However, we may label these people as 'difficult' too quickly. If we are not very careful, we will anticipate aggressive behaviour from them and we will go into meetings ready for a fight. My suggestion is that if you want to know about an individual that you are due to meet, canvass widely. Take a number of differing views on board. The perception of a single individual may well be skewed by his experience. Having listened to the opinions of others, make a promise to yourself that you will not judge that individual until you have spent some time in their presence. This will do two things; it will prevent you from subconsciously behaving in an overly aggressive or defensive manner from the outset and, it will ensure that you are attentive to the real person sitting opposite rather than waiting for them to confirm the myth surrounding them.

As an expert witness I often speak to the opposing party and find them to be courteous, well-mannered, amiable and reasonable, despite the fact that I have been warned of their rude, abrasive and oppressive behaviour. Let me cite an old example.

As a young QS, I was asked to meet an architect to settle a final account in Central London. I was told that the offices were hard to find, the Architect would keep me waiting and then would be unhelpful and rude. I was not looking forward to my day. I found the offices easily, I was ushered in to see the Architect as soon as I arrived and he was helpful. We settled the account in a few hours. On my return I quizzed my informant about his poor experience. It turned out that on the day of the meeting it was raining, the taxi dropped him off at the wrong address, and he arrived soaking wet and late. He did not have the documents he needed and he failed to reach any agreement, making the whole day a waste of his time.

You owe it to yourself to make a personal judgement on those you come into contact with; therein lie the opportunities to build worthwhile relationships, relationships that might never have blossomed if you had approached your first meeting with suspicion and fear.

8.3.2 Personalities

Take another look at the diagram of personality types in Chapter 5. Most people fall broadly into one of those quadrants. One key to working successfully with an individual is to recognise what type of person they are, and also what type of personality you are.

Look at the Ruler's personality. He will be assertive, perhaps even aggressive, but will generally want to get on with the job as quickly and efficiently as possible. Let us say that Mr Smith is a Ruler.

What type of person are you? If you are also a Ruler, the chances are you will get along well with Mr Smith. People within the same personality quadrant generally do, and this is because they already share some common ground. They understand the way each other's minds work. It is as if they were on the same wavelength. In the case of two Ruler types, the discussion may be lively or even aggressive, but agreement can ultimately be reached in most cases, because both are keen to move on.

Let us suppose, however, that while Mr Smith is a Ruler, you are an Analytic. What happens when these two personality types are in conflict?

Look at the characteristics of the Analytic. Whilst the Ruler can make decisions quickly, the Analytic needs to gather all kinds of data beforehand, weighing each fact carefully, to be sure that his decision will be right. He will not make on-the-spot decisions, because he needs time to consider every option. The Ruler sees him as a time-waster, indecisive, a 'wet blanket'. The Analytic will see the Ruler as pushy, impatient, insulting, or a 'Know-it-All'. He may be confused at how the Ruler can possibly make a decision without every last piece of the evidence being available. Ultimately, the Ruler will probably see the Analytic's apparent inability to make decisions as a sign of weakness, and will try to steamroller him into giving in.

In the Netherlands, two years ago I witnessed an example of this in person. The PM was due to go to a meeting with the client and he needed to know what extension of time was required to absorb the overrun. The planner said he would need two days to re-run the overall programme. The PM wanted a ballpark figure now. The planner said he could not give such a figure without at least a couple of hours drafting a high-level programme. The PM blew his top and sent the planner away. The PM was fuming, as was the planner. Fortunately, I had been brought in to look at the delay and I was able to say to the PM that the extension of time required was 11 weeks, but that he must explain the caveats surrounding that forecast. As a Ruler he was happy with that solution. Neither party was in the wrong; they simply did not understand each other's point of view. Their personalities clashed and clear communication suffered.

These variable behaviours are inherent within all of the different personality types, and we need to recognise our own personality group, as well as that of others, if we are to contain conflict successfully.

Individuals have a tendency to react in set ways to certain situations. These reactions are instinctive, automatic, and often negative. If you find yourself reacting negatively, examine your reaction and the effect it will have on the situation. If you believe it will have a negative impact on the situation, substitute your negative reaction with a positive one and re-run the scenario. You may find that a positive reaction offers a better outcome. We can all exercise closer control over our circumstances with a little practice. It will not be easy at first, but if you keep trying you will see differences, not only in your own behaviour but in the way that others react to you.

The next time you feel yourself beginning to react negatively, stop yourself. Ask for a break, a 'time-out', to gather your thoughts. Step back from the situation and look at it objectively. Ask yourself exactly what you want to achieve. Perhaps this will include staying calm, making your point effectively, or being able to assert your rights without losing your temper. Perhaps it will simply be to act more confidently. Whatever it is you want, just do it. Do not allow yourself to be distracted. There are three steps for staying on the right track:

1. Know what you want – have a specific goal or target.
2. Pay close attention to everyone and everything.
3. Be open-minded and flexible – consider all options.

To have a specific goal or target may seem rather obvious when facing a negotiation, but it is important not to lose sight of the real issues. Too often conflicts arise, which have little or nothing to do with the main topic of discussion. Do not allow yourself to be sidetracked. Say to your opponent:

> I understand what you're saying, and I agree it's an important issue, but we're not here to discuss that today. Let's schedule a meeting to discuss this subject, as you feel so strongly about it.

Such a statement reinforces to your opponent the fact that you are interested in what he has to say, whilst conveying to him that there would be a more suitable time to discuss it. It lets him know that you are willing to be reasonable, but also that you need to discuss the main topic now.

Paying close attention can be difficult, especially if faced with a person who is negative, complaining, pushy or hostile, or even just boring. When listening to others, do not make assumptions. These can become self-fulfilling prophecies, as we noted above. Listen instead to the way things are said. What tone of voice is used? What about the person's posture? Is it passive, or assertive? Are they attacking or defensive? Do not forget to examine your own speech and behaviour. Are there personal behaviours that might make others assume things about you? If what you are doing is ineffective or counter-productive, stop doing it.

Remember that on occasions you will find some people are easier to get along with at certain times of day. For example, let us suppose Mr Smith's secretary wants to ask for time off. She might prefer not to approach him about it in the morning, because he is always difficult to approach in the mornings. However, after a long, productive lunch with a client, he may feel less tense, and so she will know to approach him mid-afternoon. With some people, it may be the other way around. Everyone is different.

Being open minded and flexible may require practice. You may feel yourself reacting negatively to a difficult person, but try to stop yourself doing this. Try something different. Ask yourself what could have happened to make that person difficult today. They may take their frustrations out on you simply because you are there. Do not take it personally. Give them the benefit of the doubt. Say something like:

> Are you angry at me, or is something else worrying you? Do you want to talk about it?

This will help them to see that their behaviour is unacceptable, and it also signals that you are willing to listen. They may then feel that they can trust you. There are also a number of psychological techniques for building trust and rapport.

The more common ground you can find, the greater your chances of success. Look first at points on which you both agree, to establish a rapport early. If you simply concentrate on your own needs while ignoring the other person's, you will soon have a conflict situation. So how does pacing work in this situation? The first step is to listen to what your opponent has to say. Show interest, and then make a positive comment about it. Make sure the person feels that you have understood his point of view. That is the essence of pacing – making your opponent feel understood, while reducing the differences between you. It means building common ground wherever that is possible. The intention is to put yourself on the same side as your opponent. Indeed, do not think of him as an opponent at all, but as your equal, your counterpart. By stepping to his side, you automatically make it difficult for him to argue with you. Make him your friend. Win his trust and co-operation. To do this, you need effective communication.

A study at UCLA in the United States revealed some interesting facts about communication. One of the most surprising findings concludes that only 7% of communication takes place through the actual words we use. The sound of the communication, or the tone of voice used is 38%. The remainder, some 55% of what people respond to as communication, takes place visually. This includes facial expressions and body language, and these are the very aspects which, for most of us, take place on a subconscious level. They are instinctive. We do not always realise what we are doing.

Pacing is a way of creating common ground between yourself and another individual. It can be used to circumvent the conscious mind of the other person and feed directly into their subconscious. Psychologists say that using our vital signs for the rhythm is a very significant way of establishing commonality.

Most of us already pace others without even realising that we do it. Next time you are in a meeting, or in any one-to-one situation, try to be consciously aware of the things you do. You may be unconsciously sending messages to other people about yourself, so you need to be aware of how they see you and what conclusions they may be reaching about you. Most of us do this on a subconscious level. Often we do not know why we feel a certain way about someone, because we are responding on an intuitive level. Pacing is all about becoming aware of these subconscious messages, then making them work in our favour.

When you speak with someone, you tend to mirror the actions and behaviour of the other person within a short period of time. If the other person folds his arms, or crosses his legs, within 60 seconds there is a chance that you will do the same. You probably do not even realise that it is happening. Try it out on others and you will see that they take up similar body postures, adopt the same kind of facial expressions, and perhaps match your tone of voice. If you are walking along a street, they may even match your step. What this does is increase in a very fundamental way the amount of common ground between you.

There are powerful ways that this kind of pacing can work, as long as it is not done clumsily or in too obvious a way. One method is to match the person's breathing. Their subconscious picks up on this, and they respond more favourably towards you without really knowing why. Another way is to make a specific action, such as tapping a finger on the desk, each time they blink. The results may not come all at once but, over a relatively short period of time, pacing can build up a rapport with someone. They will feel that you are on their 'wavelength', that they can trust you, that they like you. They will not really know why, but you will.

You can also pace in less subtle ways. As we have said, the tone of voice used accounts for 38% of communication response. This covers not just the tone of voice but also the speed of speech, and whether it is loud or soft. Try saying the same sentence, using the same words, in different tones of voice. The words, the 7% of what people respond to, mean different things according to how they are spoken. What is the emotional content? Are the words spoken with anger? With fear? With concern? With reproach? With respect? Are they spoken in a defensive tone, or with aggression? It does make a difference.

The speed of communicating can be paced, again without the other person being consciously aware of it. If your counterpart speaks in a slow, deliberate fashion, slow down your own speech, with plenty of pauses, without making it too obvious. On the other hand, if he speaks rapidly, increase the rate at which you speak. Rapid speakers can become impatient with those who speak slowly, while the slower talkers may feel intimidated or pressured by someone whose speech is much faster.

This can also apply to the pitch and volume of speech. Pacing these is more difficult. If your counterpart has a loud voice, pacing him may lead to you shouting at each other. Make only small, subtle changes to your own behaviour when pacing in any way. If the other person has a high-pitched voice, take care that he does not interpret pacing as being parodied or ridiculed. Again, very subtle and minor adjustments to your own speech are all that are necessary. Do not go overboard with pacing, because it can backfire.

So, accepting that 7% of spoken, face-to-face communication is achieved through the words we use, we need to plan our meetings well and prepare ourselves to speak appropriately. Words, of course, assume a much greater importance when the communication is indirect, such as the written word, or when conversations take place over the telephone. In the latter case, it is the tone that becomes the most significant factor, while the written word relies upon effective punctuation and phraseology to convey the right meaning.

For the purposes of this discussion, we are considering how we manage face-to-face meetings. You may think that, if only 7% of what you are trying to communicate depends on the words you use, they are not going to be very important, but that is not at all true. The key point here is that we must listen attentively to what the other person is saying and how they are saying it. You can employ another effective technique here to build common ground, and to make your counterpart feel that he is valued and understood. It is called backtracking.

We all have a desire to be understood. When we feel that no-one listens to us, or cares about what we say, we put up barriers and cease to communicate effectively. We can overcome this feeling in the other person by backtracking. The first step in backtracking is to listen carefully. We must resist the urge to interrupt and let them finish. When they have fallen silent, repeat back to them what they have said. This confirms to them that you have listened and paid attention, also that you valued what they had to say. You then explain precisely what you have heard and what you understand them to be saying, showing that at the very least you are both in agreement with the meaning of their communication.

Anyone who has had a partner of any kind will know that there is sometimes a hidden message behind another person's words. It is for us to discover the real reason for the communication. We must look beyond the actual words used, to try to discover why the other person has phrased their sentence the way they have. It is in the nature of human beings that they do not always say exactly what they mean. Sometimes, as individuals, we may be afraid of causing offence, of being too blunt or direct, so we hedge around the real meaning of the communication, seeking signs from the other person that they have understood our unspoken message. It is under such circumstances that most misunderstandings occur. If the intent or reason for the communication is not obvious, do not make assumptions. Ask specific questions in order to clarify the meaning. Do not be afraid to ask questions such as: How do you feel? What do you mean? Why do you think this way? What is really worrying you?

Getting to the heart of any oral communication can help to eliminate conflict. Often, in a dispute or negotiation, people will use the content – the words used – as the battle ground and if they do, they will fail to see the reasons behind the words. Resolution generally comes when we discover the main intent or reason for the communication. Once clarity has been found, they often say, 'I didn't realise that was what you meant' or 'This is what I was trying to say all along.'

State your intent or objective clearly at the outset of a discussion. When your counterpart understands the reason underpinning your message, he will be more likely to listen attentively. It also increases the common ground, since at the very least you both understand what is expected or proposed.

Finally, try to avoid negative statements. Saying such things as 'You're wrong' or 'That's a stupid idea' will instantly alienate your counterpart. Try saying, 'That's an interesting viewpoint. However, have you thought about it another way?' This is much more likely to achieve results, because the other person continues to feel valued, to believe that his opinion counts for something. Choose words that build the relationship between you, even if you disagree. The seeds of success are sown when you communicate effectively and establish a rapport.

8.4   Tactical behaviour

Up to this point we have discussed the way people instinctively behave. We have examined the perceptions and personality traits that can cause misunderstandings. Clearly the avoidance of misunderstanding is critical to harmony on a project. There are, however, those people – and construction appears to have an ample supply – who deliberately behave unconscionably. These people know what is necessary to avoid conflict, but choose to behave tactically to see whether an advantage can be gained from behaving badly.

Some of the tactics are subtle and some are brutal, but avoidance of them all is necessary for a project's ultimate success. We can highlight the most common tactics and show you how to rebut them without increasing the prospect of undue conflict.

8.4.1 *Deceit*

Being economical with the truth, lying, withholding the whole truth, bluffing, misleading and misrepresentation are all aspects of human behaviour that should be controlled. As mentioned earlier in this book, if the untruth itself does not cause a conflict to arise, then certainly the discovery of the lie will cause contention.

If you believe that someone is lying, persist in asking questions and delve more deeply into their knowledge of the topic. An unprepared liar will trip himself up in due course. Where it is clear that someone has lied, draw their attention to their erroneous understanding and give your evidence in support of the facts or the truth. You must always, however, try to be tactful and give them the opportunity to save face. When they do, accept their explanation gracefully. It may be difficult, however, to trust their honesty subsequently.

An example of this behaviour occurred on a project of mine in Wales some time ago. The Contractor's site agent asked the mechanical contractor's site supervisor to divert a cold water main. In accordance with the contract, the supervisor asked for a written instruction. The Contractor's agent feigned insult at the suggestion that his word was not enough. Not wishing to insult the agent further, the water main was redirected. A month later, the QS for the sub-contractor asked for the variation to be priced and included in the interim application. The Contractor refused to pay as there was no written instruction. The agent knew very well what he was doing all the time.

Deceit makes only enemies and creates uncertainty, which is the breeding ground for serious conflict.

8.4.2 Blackmail and coercion

Used more often in construction than we would like, these two primal tendencies always cause conflict. The conflict may not arise immediately, but a person coerced into a poor decision will fight back somewhere at some time. The most obvious situations are:

- *Coercion*: 'If you do not settle this claim at the figure I have suggested, then you will never work for the company again.'
- *Blackmail*: 'You let this item go, agree that it meets the specification, and we won't have to mention that little matter last Monday.'

These examples are quite brutal, but many more subtle occurrences can be cited and will have been experienced by most professionals.

8.4.3 Bullying and duress

Strong personalities (Rulers) abound in the industry, and often these individuals have come up through the ranks to high office. As a result they may have retained a number of the habits gained when working on site.

Physically strong individuals will often threaten or hint at the use of violence, whereas in other cases, violent language can be enough to bully people into acting.

If that work isn't handed over you will suffer, believe me!

This type of duress, the fear of physical or mental trauma, is not the only type employed. Economic duress is commonplace in contractual relationships, for example:

If you refuse to accept £50 000 for that variation, you will get nothing until the final account is settled in two years' time.

When agreements are made under duress they will often fail, and if they do not, they may be challenged in the law courts. Almost without exception, this behaviour leads to conflict now and conflict tomorrow.

8.4.4 *Harassment*

Racial and sexual harassment of others obviously causes conflict, but less obvious forms of harassment produce the same results. Seeking settlement of an overdue final account, a brickwork sub-contractor wrote daily to the debtor. Eventually he had to write to the Managing Director of the debtor company and he repeated his request every day. At the end of one week, the Managing Director stopped all payments to the sub-contractor on all jobs until the matter was resolved. The brickwork sub-contractor pulled off site and issued a writ. The Contractor counter-claimed. By the time the case was due to be heard, both companies were in receivership.

Harassment, no matter how apparently justifiable, does not work. It simply causes a negative feeling and begs a hostile response. The result is conflict.

8.4.5 *Sarcasm*

Sarcasm and practical jokes, by their very nature, have a target. To be at all funny, they must hit that target, and do so in the presence of an audience, thereby humiliating one for the amusement of many.

The targets of sarcasm do not respond well to this type of attack. They may appear to take it in good part, but they seldom do. Your target will sit back and wait until an opportunity to 'get even' arises, and then will leap upon it. The result is a conflict, which is capable of escalation. Some offices and even companies have been devastated by the effects of escalating practical jokes and the poor morale caused by stinging sarcasm.

8.5 Summary

Everyone is capable of bad behaviour of one type or another. The key to avoiding or reducing conflict between people is to follow this checklist:

- Ensure that your perceptions are correct.
- Build a rapport with the people with whom you work.
- Avoid manipulating others with tactical behaviour.
- Eliminate negative behaviour in your own life and encourage your counterparts to do likewise.

If we can eliminate personal conflict between people, then we can probably manage the remaining conflicts more readily.

9 Managing Conflict

In previous chapters we have discussed why conflict occurs, how to prevent it, how to limit it and how to deal with conflictive people. To a degree we need to touch on some of these issues again because, once a conflict is underway, the same procedures can be used again to control the dispute.

9.1 Simple resolution techniques

Some conflicts over issues can be readily resolved without the parties having to become personally involved in those issues. These simple conflicts of ideas or opinions can be addressed in one of two ways:

- Fact finding
- Problem solving.

9.1.1 Fact finding

A few years ago, a well-known motor manufacturer placed an advertisement in the press, which proclaimed for a particular car:

> It will get you to London from Birmingham quicker and cheaper than the train.

The Train Operator consulted their advisers and complained about the advertisement. Their argument was that a car setting off from Birmingham, obeying the speed limits, could not get to London more quickly than the train. Furthermore, having consulted the AA, the overall costs of running a car, divided by an average mileage, gave costs per mile in excess of the rail fare. There was clearly a conflict here between the parties.

To resolve the matter, it was agreed that the car would be filled with petrol and would stand outside the railway station until the train was timetabled to

Conflicts in Construction: Avoiding, Managing, Resolving, Second Edition. Jeffery Whitfield.
© 2012 John Wiley & Sons, Ltd. Published 2012 by John Wiley & Sons, Ltd.

leave. The car would then drive, with an independent assessor, to London, obeying the rules of the road.

The day came, and the car waited for the appointed hour and set off. Some time later the car arrived in London and the manufacturers welcomed it, still wondering whether the train had beaten it. When the independent assessors checked, the train had still not arrived. Unfortunately, the engine of the train had failed, causing the train excessive delay. The conflict was now resolved. The fact was that on a given day, at a given time, the car did complete the journey more quickly than the train.

Many construction disputes are caused by an apparent lack of facts. Disputes over what did or did not happen are easily avoidable. By maintaining proper and adequate records, available to all parties, the facts can be established. Solicitors and consultants to the construction sector often waste many thousands of pounds in fees, establishing facts that ought to be known and accepted by both parties long before a conflict arises. It cannot be repeated often enough that record-keeping is the single most effective method of preventing a conflict from turning into a dispute.

As a young site surveyor, I was asked to monitor progress on a large construction site in the City of London. I was acting for the electrical sub-contractor. One day a stinging letter arrived from the Contractor, accusing my client of being in delay, inasmuch as the first fix conduits were not completed. On receipt of the letter, I walked around the building and discovered that the steel columns were awaiting their blockwork encasing. The conduits were due to be fixed to the said blockwork. To provide me with a record of this inspection I took photographs, with a newspaper in the picture too. At our next meeting, we complained to the Contractor that the blockwork was incomplete, and he again argued that it was complete. When he refused to walk the site and see for himself, we believed that he had given up his fruitless argument.

Several weeks later, at a final account meeting, we were faced with a claim from the Contractor, because his decorator could not proceed until the conduits were complete. We referred to our letter about the absent blockwork, and he referred to his letter stating that it was complete. He said, 'I don't suppose we'll ever know now. Let's split this 50–50.' At this time we produced the photographs and the programme as conclusive evidence that the blockwork was holding up decoration, not the electrical work. The Contractor blustered a little before he conceded the point entirely.

Good record-keeping, sharing programmes, exchanging letters, taking photographs and videos and maintaining computer records, are all essential when trying to discourage a potential claim or deflect a conflict. If facts can be established early in the process, then conflicts will be more manageable.

9.1.2   Problem solving

The civil courts were not really established to assist in fact-finding, and the adversarial system adopted in this country encourages each party to challenge the facts put forward by the opposition. Not only are the courts the wrong place to debate simple factual differences, but they are also poorly placed to resolve problems. The courts apply the law to the facts as they are presented to arrive at a judgement. The real time for problem solving is before the conflict becomes a legal issue. Yet all too often the Arbitrator, or Judge, is expected to find an answer to a problem beyond his scope of experience or knowledge. The parties must try to solve problems before matters progress too far and communication breaks down.

There are a number of techniques for solving problems, all of which can be addressed to aid resolution of a conflict before it becomes contentious.

- *Objective standards*: The first of these is the application of objective standards. For example, the client says that the natural wood doors are priced too highly. The Contractor disagrees and states that the doors are priced competitively. By reference to pricing books and joinery catalogues, an objective price can be obtained, and the conflict is no more. Another example is given below:

    Cableco, a sub-contractor, were complaining to the Contractor that a payment was overdue. The Contractor's accounts manager told them that the payment would be made in a week or so when his client had paid him. The sub-contractor insisted on payment in accordance with the contract. The accounts manager said that the contract between them was 'pay when paid', as were *all* sub-contracts issued by the Contractor. Cableco disagreed and, when no money was forthcoming, a writ was issued.

    Luckily, the writ arrived on the Contractor's PM's desk first. He called in Cableco, and together they reviewed the sub-contract conditions. There was no 'pay when paid' clause. By reference to the objective standard of the contract conditions, the potential dispute was defused and the sub-contractor was paid in full, the Contractor losing his prompt payment discount.

    Using objective standards, or standards acceptable to both parties, is a useful way of avoiding unnecessary conflict.

- *Brainstorming*: The second technique that can be used for problem solving is more radical and is often referred to as brainstorming. In this situation, a problem arises between the parties and there is a conflict of interests. One party seeks a particular outcome, the other party a different outcome. How can this be resolved without compromising, which often fails to satisfy either party?

    Brainstorming can often be helpful. To be successful at this, you need to free your mind of the positions that have been taken and look for new, even outlandish, solutions that may suit both parties.

    After a seminar on negotiation, given to a variety of companies, a young man approached me and asked whether I had been serious when I suggested that 'one side never had all of the weaknesses' when bargaining. I told him that such situations were much rarer than we imagined. After some thought, he outlined a problem that he thought was irresolvable. It was explained in this way.

Postal Express, a direct mailing company, had been awarded a contract to print and distribute direct NHS mail shots for a government agency to most UK homes. They were awarded the contract on the basis of 20p per delivered leaflet, and they were negotiating with Royal Mail on the mailing charges. For perfectly sound commercial reasons, Royal Mail were unable to provide delivery for 20p per leaflet. Postal Express seemed destined to suffer a loss.

Thinking back to my youth as an amateur philatelist, I remembered that our postage rates seemed high in comparison with other countries. As a wild suggestion, I asked him to contact one of the Eastern European postal services to see if they could mail them to the UK more cheaply than the Royal Mail.

It was some weeks later that Postal Express' young negotiator rang with 'double' good news. They had contacted the Malaysian post office, who had been able not only to mail the leaflets to the UK more cheaply, but who had also arranged the printing of the leaflets at a fraction of the cost of the UK printer's price.

This ability to suggest wild ideas without being criticised for their 'wackiness' is very liberating and can lead to innovative solutions. If problem solving can find an alternative to the conflicting positions, then clearly it has a place in managing conflict.

9.2   Control of conflict

If simple resolution techniques are unsuccessful, and the conflict becomes inevitable, then some deep consideration needs to be given to controlling the conflict. The methodology behind managing conflict is almost analogous to managing fire control. If we want to work in a fire free environment, what do we do to achieve this situation?

- *Prevention*: When we design the building, we use as many non-combustible and flameproof materials as possible, in order to reduce the risk of fire.
- *Precautions*: We do not sit and hope there are no fires. We take sensible precautions: no smoking, proper storage of flammable materials and the provision of sprinklers and hose reels.
- *Symptoms*: If it looks as though a combustible material is at risk of overheating and, therefore, liable to burst into flames, we cool down the material and take the heat out of the situation.
- *Alarms*: We install alarm systems to tell us that a fire has broken out and we ensure that the system is capable of identifying where the fire is. If we are notified early, we can avoid significant losses.
- *Starve the fire*: A fire needs heat, fuel and oxygen to burn, so we ensure that we starve the fire of the elements it needs to thrive. Halon gases are often used to flood the area, eliminating oxygen and curbing the fire.

- *Prevent it from spreading*: We implement measures to prevent the fire from spreading to other areas, causing widespread damage. We use fire curtains and fireproof doors for this very purpose.
- *Address the fire*: Once controlled and contained, we can fight the fire ourselves by using hose reels and fire extinguishers.
- *Call the Fire Brigade*: We are sensible enough to recognise when there is too much for us to handle and so we call in the fire brigade – we call in the professionals.

Controlling conflict can be viewed in the same way, and so we give examples of each stage of a managed conflict below.

9.3   Conflict prevention

What are some of the preventive measures that can be taken to avoid an outbreak of conflict? Well, a number of them can be found in Chapters 7 and 8. There are others, however.

Certainty is essential. In the preparation and presentation of the contract documents, we need to be absolutely clear about what we will do and what we expect from others. This is the only way that real agreement can be reached. If we are honourable and fair, then we will carry out our obligations faithfully and will not seek to find 'weasel words' or possible ambiguity in order to free us from a part of our accepted responsibility.

We should utilise proper procurement methods and rely upon appropriate and balanced contract forms. We must ensure that we vet our lists for those who are consistently contentious or conflictive. We are under no obligation to work with unhelpful contractors or sub-contractors, whose discordant voice disrupts every project they grace.

Openness will allow us to share not only the burdens of the contract but also the benefits. Full disclosure of all relevant information at the earliest possible moment will create a team spirit that seeks the common objective – a profitable and successful project.

Consideration for the other team members and their needs also assists. Many of the things we do can be done just as quickly and just as cheaply in a different way. Try to find the solution that best serves both your fellow team members and yourself.

If we remove the elements that can make conflicts happen, then we can avoid many of the conflicts that have occurred and have been repeated historically. We would be foolish, however, to use preventive measures alone.

9.3.1   Precautions

We probably all accept that to sit back and rely on preventive measures alone would be naive and so we need to take certain precautions in case of conflict.

By properly wording the contract conditions, we can ensure that provision is made for the most obvious of the conflicts arising in construction; for example:

- *Delays*: allow for the extension of the contract period where delays are beyond the control of a contractor.
- *Payment*: stipulate precise payment schedules, so that all participants know when payment is due and how it is to be calculated.
- *Losses*: permit a representative to assess the losses incurred by a contractor due to circumstances arising that are beyond his control.

Many other topics can be equally important, and the project should be analysed thoroughly at design stage to identify possible problem areas. Once identified, the documents can cover the issue appropriately, citing a fair remedy.

It would also be wise to incorporate clauses showing the procedure to be followed if a conflict should arise. The contract may include an agreement to refer disputes to a dispute Resolution Board or mediator before the matter is referred to adjudication, arbitration or to the courts. A properly worded clause could prevent many conflicts from getting out of hand.

Precautions and prevention are fine, but we must still be aware of the underlying feelings that give rise to conflict.

9.3.2 Symptoms

Watch all participants on a project carefully. Look for adverse behaviour that may suggest a problem. Are two foremen not getting along together? Perhaps one should be moved to another section. Is an argument brewing over variation prices? Get involved. Cool the argument and ensure that all discussions are rational and productive, even if it means sitting in a meeting that you would normally avoid.

Major disputes and conflicts rarely ignite without warning. Look for the smouldering issues and address them before they burst into life. By catching a conflict sooner, we can avoid a major contention later.

9.3.3 Alarms

Many construction conflicts have been trundling along for some time before they are made known to those who can resolve them. By this time the parties involved may have become entrenched in their positions, or they may be personally bitter towards one another. How do we avoid this?

We need to establish working or operational procedures that signal a conflict at its conception. The moment that two quantity surveyors fail to reach agreement on an issue, a conflict is born. To prevent it from maturing into discontent and disharmony, we need to address it immediately. By ensuring that our staff, at all levels, disclose such disagreements, we can become aware of minor conflicts before they cause great damage and incur greater cost. However, once alerted by the alarm signals, we must act.

9.3.4 Starve the conflict

Just as a fire needs to be fuelled, so does a conflict. A disagreement about the real meaning of Clause 13 is a conflict, but it is inert until someone wishes to make an issue of it.

To prevent a conflict, we can smother it, concentrate our resources on it, and let both sides know that it is in hand and that it will be resolved. Listen to the whole story, and the teller will be satisfied. It is dissatisfaction that fuels conflict.

We can also remove the fuel from the vicinity of the conflict. If it is a dispute over contract words, do not argue. The other side will not accept your position anyway. You will be contending needlessly. Send the document away to a respected, uninvolved and unbiased expert and let him decide what the words mean, then accept his important judgement. If the conflict is between people, remove one or both from the argument and take the conflict higher. Allow others who are less emotive about the issues to discuss them. After all, that is why we have managers and directors.

By starving the conflict of fuel, we are better placed to resolve the real issue and the conflict surrounding the issues dies down.

9.4 Prevent conflict from spreading

When conflict arises over one issue, we need to contain the conflict to that issue. All too often in the construction industry, a conflict occurs and causes bad feeling. That bad feeling leads to an urge to conflict, and so the cycle goes on.

Faced with a conflict, do not ignore it in the hope that it will burn itself out. It usually does not, and worse, it often ignites surrounding issues. The sub-contractor who is paid late says, 'Well, while I am complaining about that, I may as well mention…' The Contractor who receives a letter about being held up in area B, writes back saying, 'You held us up in area A and C…'

Keep the issues in a straitjacket. Concentrate on a single issue at a time. You may find that, in resolving one issue amicably, the others fall in line behind the first.

Call the parties together, restate your joint goals – a successful project – and obtain everyone's commitment to that goal. Finally, refer the dispute for resolution in an agreed form, removing it from the arena, thereby preventing it from contaminating other aspects.

Examine any conflictive project and you will find that one dispute led to another, and one sub-contractor disaffected another, and so on. Preventing a conflict from spreading makes the difference between a job with problems and a problem job.

9.4.1 Address the conflict

Once the conflict is controlled and contained, we can use our best endeavours to address it positively and co-operatively. Both parties need to address the conflict and see it as a common enemy. Conflicts never benefit anyone. Any conflict is capable of resolution. It may mean mutual agreement, compromise or total submission. The secret is finding the resolution method that is most likely to be acceptable, and most likely to succeed.

Jointly addressing a problem, as we discussed earlier, can be a way of building relationships rather than destroying them.

Do not concede defeat too easily. The next suggestion may be the one that will bring agreement. You will never find out if you do not persevere. Make every effort to resolve the matter between yourselves. Try not to rely on the third parties on a regular basis. Use the negotiation techniques in Chapter 10.

If, having done all of this, you are still unable to resolve matters, look further afield.

9.5   Call the professionals

When a conflict or dispute is out of our control, then we need to act rapidly to avoid a total loss situation arising. Our first priority must be to call in those who can eliminate the problem quickly. There is no shame in having to use a third party to resolve a dispute. It is commonplace in the industry, the outside world, and even in marriages.

An outside person will address the conflict from a new, independent and unbiased standpoint. They will be uninvolved in the circumstances surrounding the dispute and will, as a result, remain unemotional. A third party may see things that the first two parties have overlooked. The third party will also be an expert in these matters, and his experience will enable him to identify the real problems and the real issues. An astute third party will recognise the respective positions of the parties, and look beyond them to the commonality of the parties' interests.

A fresh approach to an old problem is often a key to a solution. In 1992, I was asked to act as an expert witness on a dispute that was already eight years old. The dispute involved a mechanical contractor who had, quite clearly, been disrupted in the carrying out of the works. His claim was a simple one. The plumbers who fitted the heating in the flats were unable to proceed regularly and diligently with the works. The tenants, the authority and the gas company had all hindered them. I was representing the authority, and had to concede that my client was likely to be liable for the delay.

The Contractor wanted around £100 000 for loss of productivity and prolongation. The authority had refused to settle the matter and the dispute rumbled on.

When I was called in to look at the dispute, I ignored all of the rhetoric and the conflictive letters, referring back instead to the actual cause of the dispute, the disrupted plumbers. I noted immediately that the plumbers had indeed visited each flat up to seven times, rather than the single visit planned at tender stage. I also noted that the plumbers were self-employed, and that they were being paid on results. Each pair of plumbers received £200 per flat, regardless of how slowly or quickly the flat was completed. This meant that it was the plumber who bore the loss of productivity and not the Contractor. It quickly became clear that the Contractor had not suffered financially at all as a result of lost productivity. His losses had clearly accrued elsewhere. On further investigation, it was discovered that, before the plumbers started on site, the Contractor agreed £200 per flat with them, whereas he would recover only £180 per flat in his tender sum.

A new pair of eyes spotted an obvious problem that the involved eyes had missed, and the case was settled without a hearing. So, where it is necessary, call in a professional, but do it quickly.

One thing that we should avoid is employing professionals who ramp up the conflict rather than assist in the resolution. The best and most economic use of Claims Consultants is to bring them in early and allow them to fact find, to assemble your factual records and advise you on your position once the records are gathered. You would not say to a garage: 'Here's my car, see if you can find anything wrong with it and repair it,' because ultimately you would probably have to sell the car to pay the bill. Yet that is exactly what many parties do with Claims Consultants. They give them the files and ask, 'Can I make a claim here?' The answer 'No!' will be rare indeed.

9.6   Summary

The term 'managing conflict' must include avoidance, reduction and control of conflict, as discussed in the three last chapters. Conflict can be avoided, reduced and managed if we use the techniques available.

We need to commit ourselves fully and honestly to dealing with people with understanding, taking time to identify conflicts and asserting ourselves to addressing problems quickly. If for any reason we should fail, then the next chapter will help us redeem the situation, but remember, prevention is not only better than a cure, it is also cheaper.

10 Informal Dispute Resolution

Some conflicts are simply unavoidable. Proper management of the conflict will ease the impact it has on the construction process, but resolution must follow quickly. Conflicts cannot be allowed to fester or lie dormant, they must be addressed. An old conflict is a bitter conflict.

How, then, do we set about resolving conflicts? The answer is two-fold – either informally or formally. This chapter is dedicated to the informal processes.

Informal resolution has two major avenues down which disputants may travel. They are:

1. Two-party negotiation
2. Two-party negotiation with assistance.

We will examine both in some detail and show what each has to offer to the conflicting parties.

10.1 Negotiation

If it is possible for the two parties to discuss the problems amicably and without antagonism, then negotiation should be the first choice for the resolution of a dispute. The positive benefits of negotiation are:

- Negotiation is inexpensive.
- Negotiation maintains relationships.

On the negative side, the pitfalls of a negotiation could prove to be:

1. Negotiations fail after a long and protracted period of discussion, because they are not binding.
2. The informality will permit negotiations to raise surprise issues or irrelevant points.

Conflicts in Construction: Avoiding, Managing, Resolving, Second Edition. Jeffery Whitfield.
© 2012 John Wiley & Sons, Ltd. Published 2012 by John Wiley & Sons, Ltd.

If the negative points can be eliminated – and they can – then negotiations can be worthwhile and successful.

The way to deal with 1) above, is simply to establish a timetable for the negotiation. Set aside a number of hours or days for the negotiation, and limit the duration of the discussions. You might agree that if matters cannot be resolved in five days, then both parties will call in an adjudicator. There are surprising benefits to negotiating within a time limit. It concentrates the negotiators' minds. I guarantee that on the afternoon of day four, they will be avidly trying to settle the dispute, to avoid having wasted five precious days.

To deal with 2) above, we can set an agenda. We can agree to limit the points in discussion and outlaw time-wasting discussion on irrelevant issues. The benefits of so doing are not just to avoid acrimonious debate, but it also helps the parties by limiting the issues in dispute. When written down in agenda form, some issues will look decidedly petty or silly, and will be withdrawn.

Some issues will disappear, because one side or the other will not consider their value to be great enough to spend time in discussion. These items will be agreed or withdrawn as a result. The agenda will then highlight the real issues requiring serious consideration. Again, minds will be focused on specific problems such as:

Why did you fix the ceiling on the third floor east wing four days late?

rather than trying to answer a question, which is too general, such as:

Why were you consistently late fixing your ceilings?

Precision in questions, and on agendas, will encourage precision in answers. Once the disadvantages of negotiation are overcome, we have a very powerful method of overcoming conflict at our fingertips, assuming we know how to negotiate effectively. Some guidelines for effective negotiations follow, so that inexperienced negotiators may learn the techniques and experienced negotiations may refresh their memories.

I once presented a seminar in the Midlands on the topic Effective Negotiation, after which the Commercial Director of a national construction company approached me and confided that he had not intended to attend the seminar because, in his view, negotiation was simply a matter of compromise. I explained that an effective negotiation was a sure way of avoiding the need to compromise.

To view the process of negotiation as simply being a form of compromise is both inaccurate and inadequate. However, whilst negotiation is neither compromise nor diplomacy, it does encompass elements of both.

The first thing to remember about negotiation is that, just as in conflict, it takes place between people. There is a natural tendency to view a negotiation as being between two companies. This approach denies the basic truth behind all negotiations, which is that negotiation should be all about people working together. My definition of negotiation is a simple one. It is a procedure where two or more people with diverse interests and opinions meet together to agree how they will achieve a common purpose.

There are three contingent parts to any negotiation – the people, how well they prepared, and the way the negotiations are conducted.

10.1.1   People

To be an effective negotiator, the first thing we must do is prepare ourselves. To present our case professionally and convincingly, we must first feel sure of ourselves and our position. We should be positive in our attitude and we should develop a positive self-image. So how do we know whether we have a positive self-image? We can use this simple test. A person with a negative self-image will be inclined to say, 'That can't be done' or 'I can't do that,' whereas a person who has a positive self-image will be more inclined to say, 'I can do that' and 'I will do that.' If you have a negative or pessimistic viewpoint, this will become apparent to the other parties in the negotiation, and it may reduce your own negotiating team's confidence in you as well as projecting to the opposition that you do not expect to win. In preparing yourself for a negotiation, you must not only have a desire to win, but a burning desire to win. A positive attitude that anticipates success will communicate itself, non-verbally, to the opposing parties.

The other important people involved in a negotiation are the opposing party, and efforts should be made to establish a rapport with them very early in the process, preferably before the negotiations begin. Skills should be developed that enable us to build a rapport with the other side, as they are more likely to deal honestly with those whom they trust. There are a great many techniques that we can use to establish camaraderie, and these are outlined earlier in this book. The three that I would suggest as the most relevant in terms of negotiation are smile, use their first names often, and ask questions, listening for the answers.

In a negotiation, we must be both confident about our own position and capable of winning the other party's assent. This is far more likely to happen if the personal barriers between the parties have been removed.

10.1.2   Preparation

Without proper preparation a negotiation will fail to achieve its stated purpose. A form of agreement may well result, but one side, or both sides, will be discontented. Great sportsmen can usually accept defeat, but usually find it harder to accept a poor performance. It is the same with negotiators. If we feel that we have not done our best, we will be unhappy. The best way to avoid this unhappiness is to prepare. The first rule of negotiating is:

No negotiation without preparation.

If you have not had time to prepare, simply delay the negotiation. Do not proceed without arming yourself with all of the facts. The day you do so, will be the day that the other side have all of the facts, and the day you lose.

Preparation will be both general and specific, and a checklist for general preparation would be as follows:

- What do we know about the company?
- What do we know about their negotiators?
- Can they make their own deals? Can they make deals that will stick?

Investigate, as far as you can, the company with whom you are negotiating. Understand their culture, their aims and how they are placed financially. Just like people, companies have weaknesses and strengths.

When representing a client in negotiations, we were trying to show that the Contractor had caused some of his own delays. I represented a local authority, which was willing to grant some extensions of time but not all that were requested. The Managing Director of the contracting company rejected the idea that his company had delayed handover of a number of dwellings because of poor finishing work. It was his site agent's word against the clerk of works' word, and quite naturally he believed his own agent.

During the discussions, I produced a list of previous contracts carried out by the Contractor, which showed that, in each case, they raced ahead at the beginning, yet failed to complete on time. With this history came a letter from a housing association, who had used the Contractor, which read in part:

> …and so, although they do tend to have problems in finishing dwellings off, we would certainly use them again.

The Managing Director conceded the point and an acceptable extension of time was granted.

Knowing about the company and its history or reputation is valuable in negotiation.

Of equal importance is our knowledge of the individuals who represent the company. As well as seeking their company objectives, each individual will be pursuing his own personal objectives. Perhaps he is seeking the praise a good settlement will bring, or promotion, or perhaps he simply wants to get home early and set off on holiday before the rush-hour traffic builds up. By understanding his personal needs, you may be able to satisfy those without conceding anything to his company. Have you ever got a better deal than you should have, because the other side's mind was elsewhere? The answer will probably be yes, so use this advantage and, just before the rush hour, suggest a solution. You may be surprised at the relief you see on his face.

The second great rule of negotiation is:

> Knowledge is power. Gain as much knowledge as you can and give away as little as you can.

This knowledge should help you to identify the answer to the third question, 'Can they make the deal?' There are a great many reasons why people cannot settle disputes themselves. Sometimes their superiors set limits. They may have financial limits imposed by a bank or their accountants. The individual may lack the self-confidence to conclude the deal without referring back to his superiors. Whatever the case, the result is the same – a long negotiation, no resolution and possibly starting again on another day, with another person. Do not be afraid to ask this question:

> If we reach agreement today, are you in a position to honour that agreement without referring back to others?

If the answer is yes, then that is fine. If the answer is no, perhaps you should consider speaking to someone with more authority.

The remainder of our preparation for a negotiation should centre on the specific conflict that is the purpose of the negotiation. One of the greatest failings of negotiators is that they fail to identify the specific purpose for the negotiation and, perhaps more frequently, when they do identify the purpose, they allow themselves to be blown off course during the negotiation. A negotiation will only be successful when a commitment has been made to a specific and defined purpose or objective, and that specific purpose or objective is kept at the forefront of your mind during the discussions. The art of good preparation is to understand your purpose and to anticipate the tactics of the other party, who will try to deflect you from it.

Some of the questions that we need to ask ourselves before the negotiation, and to answer to the best of our ability, are: What do we want? What can we afford? And, what is the probable outcome? The questions that we need to ask about the opposing party include: What are they likely to offer? What are they able to pay? Under what terms would they expect us to settle? A great deal of statistical analysis can be done to assist in establishing the answers to these questions, but this analysis is beyond the scope of this book.

One simple way of establishing how prepared we are for a negotiation is to have a practice negotiation within our organisation. To enable us to do this, we usually require someone highly placed in the organisation to act as if they were the conflicting party. If we use an individual who is junior to us in authority, they will have a tendency to agree with us. They will also lack the experience to anticipate and put forward the reactions and comments likely to emanate from an opposing party. A robust and honest pre-negotiation should identify the weaknesses in your case, enabling you to prepare a strategy for their defence. The pre-negotiation should also identify your strengths, so that these can be exploited as far as is possible.

As with most areas of life, knowledge is power, and therefore the greater the knowledge you have about your own case and the opposition's, the more powerful you will be. General knowledge is useful, but specific knowledge can swing a conflict negotiation completely in your direction.

In the early 1960s, in a televised, face-to-face, pre-election debate, John F. Kennedy used some specific information about Richard Nixon to win the debate. Richard Nixon had been ill in the run up to the presidential campaign, and required a great deal of make-up to appear before the cameras. During the early stages of the debate he perspired heavily, and the make-up began to slip down his face and hung precariously around his jaw and under his chin. This was not noticeable face-on to the cameras, but to John F. Kennedy, who was looking at Richard Nixon's profile, it was very obvious. It was clear to John F. Kennedy that Richard Nixon was not only feeling very uncomfortable but also that he was unable to turn his face in profile to the camera. Armed with this knowledge, John F. Kennedy spent the remainder of the programme challenging Richard Nixon to look him in the eye, and continually turned to look at Nixon while asking and answering questions rather than looking into the camera. Nixon could do no more than stare intently at the camera for fear of the make-up dislodging altogether. A poll taken after the debate showed that most people considered that John F. Kennedy won the debate hands down, and that Richard Nixon had seemed lack-lustre, unconvincing and nervous.

This of course is an extreme example, but it does reflect the way that specific information can be used to your advantage in a negotiation.

Preparation not only helps you to promote and defend your negotiating stance, but also gives you the confidence to deflect tactics when they are used against you. These tactics will be discussed later.

10.1.3 Negotiation: Building rapport

When resolving a conflict by negotiation, you should try to establish a rapport. This necessary step can be achieved easily; for example, by sitting on the same side of the negotiating table as your opposite number. Sitting on the opposite side of the table can put a psychological barrier between the parties, which may then need to be overcome. It is therefore preferable to avoid unnecessary hindrances, if possible. If there are two negotiating teams involved, it can be useful to sit professionals together in their disciplines, rather than have your team on one side and the other team opposite. Where this is not possible, it is sometimes preferable to rearrange the chairs to give the impression of a round table discussion, rather than one team facing the other.

Recently, in France and in Oman, I suggested this approach in two very different meetings. In both cases, there was an increased amount of agreement between the disciplines sitting next to one another, often to the annoyance of their team leaders. Nonetheless, the meetings ended in substantive agreements, which neither side felt like celebrating, which often signals a good outcome.

Another way to establish rapport is to make sure that your first questions are specifically designed to elicit yes answers. It is well-known that in a negotiating situation, once the momentum of agreement is rolling, then the parties are reluctant to break the momentum by saying no, unless it is absolutely necessary. Careful consideration needs to be given to the first few questions that are asked, to ensure that they are non-contentious and the other party can agree without making any concessions. Designing questions that are bound to receive a yes answer is not easy and takes some practice, but it is definitely worthwhile.

If your opposite number is an experienced tactical negotiator, he will appreciate that you are not likely to give up your position easily and move towards his. He may therefore use a number of tactics specifically designed to move you from your position into no-man's-land.

Once you are in no-man's-land, and you have lost sight of your destination, it is much easier for him to persuade you to follow him. It is therefore of paramount importance that in any negotiation you keep your purpose firmly in mind and do not allow yourself to be diverted from that goal by the tactics of the opposition. There are a number of obvious tactics, which we have all seen used, but may have failed to recognise as tactics, and these are discussed next. Before we move on to discuss them, remember that all tactics are designed to elicit concessions from you without any in return. So always follow the third rule of negotiating:

Never give a concession without receiving one in return.

10.1.4   Negotiation: Tactics

If conflicts are going to be resolved, we must avoid tactical behaviour. We need to build a relationship, or perhaps rebuild one that has been damaged or even demolished. This is only done by honest, ethical and assertive negotiation. The tactics used below have a single purpose – to prevent you from attaining your goal, a fair resolution to an unavoidable conflict.

Study the tactics and then, when you recognise them being used against you, rebut them or simply tell your fellow negotiator that you prefer open and honest negotiations. Once rumbled, most tactical negotiators will settle disputes in a principled way.

- *The wince*: This tactic pressurises you into reconsidering your offer without actually verbally rejecting it. For example, you need more time to carry out a variation, the client asks how long you need and you tell him. He winces and takes a sharp intake of breath. You now feel uncomfortable and you take your eye off your goal. At this particular moment, you may find yourself reducing the requested period without any concession whatsoever from the other side. In a sales training manual for a major motor retailer, I read the quotation: 'A good wince can make you thousands of pounds a year in commission.' This is a cynical and well-used tactic. You must ignore it.
- *Silence*: Like the wince, this is meant to put pressure on you, by causing you to be uncomfortable. You make your opening statement and the other party just sits silently, saying nothing. Because most people dislike long silences and find them embarrassing and uncomfortable, you are tempted to fill the silence. Generally speaking, amateur negotiators are so desperate to end the

silence that they will fill it with a concession. Once again, the opposing party has extracted a concession because you took your eye off the goal, and he has had to concede nothing. Practice sitting in silence, learn to be comfortable with silence in negotiations. If you feel that you must break the long silence because you feel embarrassed or uncomfortable, break the silence with a repetition of your first offer. Do not offer any concessions. Quite often you will he able to restate your position and then sit silently yourself waiting for a response, thus transferring the pressure to your fellow negotiator.

- *Lack of authority*: You are in the middle of a negotiation and you offer to make a concession if the other side makes a concession in your favour. The person opposite accepts your concession, but says that he is unable to make the concession in your favour because he has not got that authority. This tactic can be used in two ways: firstly to block negotiations that are not going in his favour, and secondly to enable him to refer back to his superior. He can then use his superior as an excuse for not giving a concession in return. If this happens, you should either insist on speaking to the superior so that the negotiations can be concluded satisfactorily, or you should withdraw your concession. On too many occasions, we sympathise with the awkward position of our fellow negotiator by allowing our concession to stand when we have received nothing in return. This tactic is commonly used deliberately, and is an abuse of your good nature. Beware of it.

- *Red herring*: This is a blatant attempt to move you away from your purpose and to draw your attention to a false premise. An example of this is where a sales person will offer to give you a discount of 20% from the price list. Often our next question is whether that discount compares with what our competitors are being given, and we then set about negotiating the discount percentage rather than the price itself, which was our former intention. The salesman has then succeeded in moving us away from negotiating the actual price, to negotiating a discount on a price list that he has originated. This is a very common tactic, and a very successful one for sales people. The way to counter this tactic is to refuse to take your eye off your goal and when he offers 20% off their price, list ask the question: 'How does that relate to my offer of…?' and then insist on returning to the negotiation of the price rather than allow yourself to be misled by a spurious discount percentage.

- *Better offer from others*: This is a common ploy used by those offering work to those tendering or quoting for work. It is, in fact, a form of bluff, where the person offering the work explains that whilst your price is otherwise satisfactory, and whilst he would he happy to use you to carry out the work, unfortunately he has a better offer or a lower offer from another party. He then suggests that you reduce your price in order to become the lowest price tenderer so that he can accept your offer. Sometimes the person offering the work will tell you the price he expects from you, thereby putting you under pressure to reduce your offer to that level. On other occasions, they will not even give you a target price and will leave you to make that decision for yourself, hoping that you will reduce your price by even more than they expect in order to win the work. Once again this is a pressurising tactic, which is intended to change your goal from achieving an attractive price for the work to the new goal of winning the work at the lowest price he can get.

- *Trial ballooning*: We mentioned earlier that knowledge is power, and one method used by professional negotiators to find out your best price, is the tactic of trial ballooning. If, for example, you are seeking £150 000 to settle the conflict, but you are prepared to negotiate, your opponent will make a suggestion along the following lines: 'Could we reach a deal if I were to offer you £135 000? Is that an acceptable figure?' If your response to this question is: 'Yes, we can live with that figure', then he may well say: 'We thought £120 000 was a reasonable figure, perhaps we could split the difference and agree £127 500.' Effectively, what he has done at one fell swoop is to reduce your real asking price significantly without offering any concession in return. A number of other similar tactics are used by professional negotiators to weaken your position, by forcing you to disclose information to which only you should be privy.

Tactics will only deflect those who are willing to be deflected, and all tactics can be avoided by having a strong determination to get what you want from the negotiation and keeping that purpose at the forefront of your mind at all times.

10.1.5   *Negotiation: Agreement*

There are many ways of reaching an agreement and resolving a conflict. After a seminar on an unrelated topic, a delegate from a sub-contract organisation approached me and made the statement that he did not believe in negotiation, because it was just an excuse for the Contractor to reduce your price, thereby forcing you to compromise. I said that I considered that negotiation was a process which, if used properly, should prevent the sub-contractor from having to concede unilaterally to his employer. In fact, compromise is only one method of reaching agreement in a negotiation.

The first accepted method of reaching agreement is the use of the proposal and the counter-proposal. This is sometimes referred to as the thesis and hypothesis method. In this process, one side makes a proposal that represents his most satisfactory outcome, the other party responds by stating his position that may also be his most satisfactory outcome. By a further series of proposals and counter-proposals, the parties ultimately reach agreement at a level that is perhaps less than they would ideally have liked, but is within their range of options.

The second way of reaching agreement is by brain-storming. This requires no compromise and is based on the theory that a negotiation can be a 'win–win' situation. An example of this is illustrated by something that happened to me some time ago. I came into the house to find two of my children arguing over who should occupy the dining room. The oldest child was arguing that he needed peace and quiet to do his homework. My second child was arguing that he wanted to listen to some heavy metal music, and was only able to do so in the dining room where the hi-fi system was situated. It appeared to them on the surface that these two needs were incompatible, and that one person would have to give way and leave the room as no compromise could be acceptable. I heard the argument continuing and walked into the room. I listened to the arguments for both sides and, without a word, took the headphones from the cupboard, plugged them in, and placed them on the ears of my second child. Thus, without compromise, both children were satisfied.

Of course, compromise is sometimes necessary in order to reach an agreement in a negotiation. However, it should be remembered that if we negotiate properly, utilising the principles cited earlier, then we will be in a stronger position when it comes to a compromise, and so we should receive a better deal. Compromise is often very unsatisfactory and often leaves both parties unsatisfied, and so therefore every effort should he made to resolve the matter by some other means, or at least by making the opposition feel good about the deal he has achieved, otherwise this conflict may be resolved at a cost to future harmonious relationships.

There are some cases where you will find that you have no strength whatsoever in reaching a deal. Whilst this is very rare, it does occasionally happen that you have absolutely nothing to offer and yet you want something in return. In these circumstances, we have to resort to the final option in reaching agreement, which is to throw yourself onto the humanity of your opponent. Surprisingly, this is a relatively successful way of reaching an agreement where you have seemingly little bargaining power. It is my experience that the majority are magnanimous in victory and, having won the argument on principle, they are prepared to be reasonably generous in the settlement stage.

10.1.6   Negotiation: Settlement

When discussions are complete, and an agreement has been reached, it is important to commit to writing the result of the negotiation. The agreement should be written in clear and concise form and should accurately reflect the agreement reached. Before leaving the negotiating room, it should be examined by both parties for accuracy and clarity of intent. Having been agreed, the document should then be copied and distributed, so that when the final copy is delivered, all parties have something against which to compare it.

It often happens that at this stage one of the more cynical negotiating tactics is brought into play. In negotiating terms, it is called nibbling. This is the name given to attempts at improving the deal to your benefit after the negotiation is concluded. Often this takes the form of a discount, which is written into the typed agreement, but which was never agreed. It can also take the form of extended payment terms, which are far beyond the payment terms anticipated at the time of agreement. It is therefore essential that, once an agreement has been reached, the terms and conditions applicable to that agreement are also agreed. This brings us to negotiating rule number four, which is:

Always get your agreements in writing – written deals stick.

Negotiations are very complex activities, which involve a whole range of human emotions and responses, and therefore they need to be carefully planned. Preparation is everything, so be prepared to make your case confidently and robustly, whilst readying yourself for the inevitable tactics and counter-arguments of the other side.

In summary, there are four negotiating rules, which must always be observed:

1. no negotiation without preparation;
2. knowledge is power – get it and keep it;
3. never give a concession without one in return;
4. always get agreements in writing.

Always remember that if the other party did not want to resolve the conflict by negotiation then you would not be there, so be firm and purposeful and you will rarely leave a negotiation without having given the other side every opportunity to resolve the conflict and proceed harmoniously.

10.2 Alternative Dispute Resolution (ADR)

When a conflict can no longer be resolved by the parties alone, a third party needs to be brought into the fray. This other party will have one of two roles to play, either:

1. *Adjudicator*: deciding on the issue himself;
2. *Mediator*: helping the parties to decide on the issue.

We deal with the adjudication role in the next chapter. Within this chapter, an informal resolution of conflicts, we concentrate on the process of Alternative Dispute Resolution (ADR).

The 'Alternative' in ADR refers to another way of settling disputes beyond litigation and arbitration. The key to this method of resolving disputes is always to ask the question:

What is it that will help the parties resolve the conflict?

This provides an agenda for the resolution process that excludes laying the blame, digging up old arguments and criticising the behaviour of others. ADR has many of the advantages of negotiation, together with some advantages of its own, namely:

- cost effective form of resolution;
- time limited, often complete in a single day;
- a mediator to calm things down;
- an agenda to identify issues in conflict;
- often written statements, outlining each case.

There are other advantages, and these will usually derive from one of the above. The main advantage, as far as conflict resolution is concerned, is that it is non-adversarial.

The ADR process can be either reactive or productive. It can be in place preventing disputes and conflicts from arising as easily as it can be in place for resolving those that have arisen. On major projects overseas the parties are now employing professionals as Dispute Resolution Agents (DRAs).

10.2.1   Dispute Resolution Agents, dispute resolution boards

The role of the DRA is defined by the parties, but his main aim is to resolve conflicts as they arise and prevent them from escalating. To do this, the DRA may be permanently on site, constantly walking the project, watching progress and recording it. There may even be a formal walk around with the major players each week, to determine what progress is being made and investigating any slow areas that may cause concern later.

The DRA is also available on site for discontented contractors. They will try to resolve their differences with their customer or supplier but, failing that, the DRA will be there to listen to the arguments.

A formalised mediation can be held very quickly, without impacting on the progress of the project, and as a result the costs of a dispute are lessened. As was discussed earlier, conflicts that are left to be resolved later often cause further conflict to arise elsewhere. Prompt action by the parties and the early mediation of a dispute prevents this time-lapse conflict from arising.

The proactive element is relatively new, even in ADR terms, and will continue to develop as clients and constructors catch the vision of a conflict-free project.

Dispute Resolution Boards work in much the same way, but are constituted of a number of professionals and/or experts.

10.2.2   Mediation and conciliation

These assisted negotiation techniques are both intended to reconcile those in conflict, by using the good offices of independent and impartial professionals who are skilled in the techniques and experienced in construction or engineering.

The mediator or conciliator has, as his main goal, the resolution of the conflict by the parties to achieve a practical and commercial remedy. To do this, he will not restrict himself to the legal issues, or even to the technical issues, but he will listen to all of the parties' contentions.

The mediator or conciliator will try to prompt the parties to suggest remedies that are mutually acceptable. He may even make suggestions himself, as to the reasonable and just settlements available. However it is done, the mediation or conciliation will attempt to meet the real business needs of the parties, and these must include maintaining an ongoing relationship.

Over the years, I have had many articles published in the technical press, and the response to each was limited to a letter or two, usually of disagreement. However, in June 1988, before ADR was commonplace in the UK, my article on ADR in a construction journal reaped 208 letters of enquiry. Now we see every major consultancy and all national solicitors have ADR services in their brochures. This has been made even more popular by the Pre Trial Protocol in the UK, which demands that mediation should have been tried before a judge is troubled with a full hearing.

Conciliation is well-known to all of us who take an interest in Trade Union disputes. Almost every sizeable labour dispute is referred to ACAS for conciliation. Conciliation within ADR has many similar characteristics, for example:

- The conciliator is investigative.
- He correlates the facts.
- He tries to reconcile the opposing views.
- He prompts the parties to propose settlement offers.
- He highlights the possible consequences of failing to settle, indicating strong and weak points in the cases of both parties.
- He does not propose his own settlement position.

Mediation is a broader approach, which also allows the mediator to propose settlement terms of his own.

The clear difference between these techniques and formal adjudication is the lack of legal procedure. This is an advantage in encouraging the parties to come to mediation, because the hearings are usually without prejudice, but it does have disadvantages too. Perhaps the most difficult of these disadvantages is the lack of a legally binding result. It is up to the parties to a conflict to act in good faith and honourably uphold the settlement terms of the hearing.

Whilst these procedures are informal, the parties can give them structure, which is akin to the formal processes, in order to formalise them. These formalities may include document-only hearings or oral evidence in the presence of both parties. Flexibility and the desire to meet the needs of the conflicting parties will be the deciding factor in whether ADR succeeds as a remedy or whether it fails.

Mini Trials are yet another form of alternative dispute resolution procedure. A neutral sits on a panel with a senior executive from each party, and a hearing is arranged. At the hearing the parties are represented, and each party offers its case, referring to the 'core bundle' of documents. The panel may ask questions, but otherwise there is no cross-examination as such.

After the hearing, the parties will try to settle the dispute, using the neutral as a mediator or conciliator as necessary. After an agreement is reached, a joint written statement is prepared and is signed by the parties. The neutral witnesses the signatures and the agreement is often agreed to be legally binding.

Such is the success of the Mini Trial in the USA, that many eminent retired judges will sit as neutrals for parties in dispute.

Whether the negotiations are assisted or unassisted, some will not succeed. There will be times when a party to a conflict will be insincere in his intentions to settle. Some will cynically use negotiations and ADR to defer settlement of a dispute. In these instances, formal dispute resolution will be necessary, utilising either arbitration or litigation.

11 Formal Dispute Resolution

The aim of negotiation and mediation is to bring the parties together to find a joint solution that is mutually acceptable. Formal dispute procedures merely try to find the proper solution, whether or not it suits the parties. This is one reason why people find arbitral and legal processes so unsatisfactory. On many occasions, winners and losers have left a hearing frustrated because their case, which seemed so simple, now appears too complex for them to follow.

Whilst the formal processes are adversarial, and so pit the parties one against the other, they do follow a standard conflict resolution pattern if examined properly.

11.1 General principles of dispute management

11.1.1 Avoidance

In order to avoid unnecessary legal conflict, the courts insist that the case is stated clearly, and that the defence is properly submitted. To avoid confusion, both sides are permitted to ask for further and better particulars. Only when the courts are sure that the issues are properly defined, will they allow the case to proceed.

To assist in avoiding legal conflict, and to prevent the congestion of an overstretched court system, lawyers are encouraged to be realistic in their aspirations for their clients. If a defendant pays a sum into court, then the plaintiff may accept that sum, claim his costs and the case is settled. If the payment into court is ignored and the plaintiff wins but is awarded less than the amount in court, then generally the plaintiff pays all of the defendant's costs after the date of the payment into court. The same type of procedure is also available in arbitration, and both jurisdictions allow offer letters, known as Calderbank letters. These measures are made available so that a legal conflict, in a full hearing, can be avoided.

Conflicts in Construction: Avoiding, Managing, Resolving, Second Edition. Jeffery Whitfield.
© 2012 John Wiley & Sons, Ltd. Published 2012 by John Wiley & Sons, Ltd.

11.1.2 Reduction

If the case is not settled and is listed to run, then orders are made that should reduce the amount of conflict and prevent the conflict from accelerating.

Firstly, once pleaded, the case as stated is the case that will be heard, unless permission to amend is obtained from the Judge. So after the pleadings are made and the defence is entered, the dispute is frozen.

Secondly, the courts and arbitrators usually allow full disclosure of documents to both sides. This means that all relevant documentation in the plaintiff's files is disclosed for the defendant's inspection, and vice versa. This helps both sides to assess the validity of the case against them and to consider settling the matter by a further payment into court.

A third catalyst to conflict reduction is the order to produce a 'Scott Schedule'. This is a spreadsheet-style document, which explains the dispute or differences between the parties in tabular format. When properly completed, with both parties' cases stated, the schedule highlights the main areas of contention. It is not unusual at this stage to discover that there is more agreement on the issues than formerly thought. Clearly, this can help to reduce the amount of argument in court, and may even instigate settlement talks.

Fourthly, the courts and arbitrators often order the use of independent expert witnesses. These experts are expected to act in an honest and unbiased way to assess their clients' claims. They also comment on the opposition's case. The clear intention is that the independent experts, both from similar professional backgrounds, will arrive at similar answers, thus narrowing the issues and reducing, or eliminating, the disputed issues.

11.1.3 Control and management

The control and management of a legal conflict is exercised by use of the Orders of the Court. The Court rules govern the procedure to be followed by all parties, thus preventing disputes as to how the matter should proceed. By directions of the Judge, the timetable is set, and he has the power to enforce the timetable. If one party is dilatory in their submissions, the Judge can issue an 'unless' order. Then, unless the documents are provided by a given date, the plaintiff's right to recover is extinguished and the defendant may lay claim to his counter-claims, expenses and costs. Interlocutory hearings are held on a regular basis, and all the parties are copied in on all correspondence to prevent contention arising due to ignorance of the procedures.

11.1.4 Resolution

Having got this far, the conflict has almost reached its conclusion. In order to ensure a fair resolution for all involved, there are procedural rules that ensure fair play. All witnesses of fact are examined by both sides, as are expert witnesses. Having heard the submissions of each party, and all of the witnesses, the Judge has to apply the law in a reasoned and explicable way. The judgement handed down by the Judge is binding in law, and both parties have to adhere to its instructions.

In arbitration, it is the arbitrator who decides the issue and who makes an award, which is equally legally binding.

Only an appeal to a higher court will allow the conflict to continue beyond this, and appeals are both expensive and time consuming. I have on my desk a recent appeal decision on a case first decided upon in 2001, over ten years ago.

As was discussed briefly earlier, our English law-based legal system is adversarial. This does not augur well for conflict reduction. Each side is encouraged to find as much damaging material as possible about the other's case. This will lead to a widening of the gulf between the parties, making settlement less likely. A further negative effect on settlement is provided by the breakdown of communication between the parties as they begin to speak only through their solicitors. If the parties do speak directly, they are usually accompanied by their respective lawyers, and counselled to be careful not to give any indication of weakness. This usually serves to prevent them from giving any indication of compromise either.

Clearly, in a book that attempts to reduce conflict, it must be pointed out that litigation and its clone-like twin, arbitration, are the least admirable ways of concluding a conflict. More often than not, both sides lose as the costs of management time, witness costs, expert fees and legal costs rise. Even winners do not collect the whole of their costs. Around 60% is often quoted as a reasonable expectation of recovery of your own costs. With solicitors at well over £200 per hour, and good counsel in excess of £300 per hour, it does not take long for the costs to surpass the value of a medium-sized conflict, and then the issue becomes not who was right or wrong, but rather who picks up the bill for the costs. Many a dispute that was capable of settlement on the issues has had to run to a full hearing, because the plaintiff or defendant needed not only to win the issue but also to win his costs.

11.2   Arbitration and Litigation

There are a great many learned textbooks on the procedure and practice of Arbitration and Litigation. Mostly they run to many hundreds of pages, and it is to these learned tomes that I would refer you for a full understanding of these processes. However, for completeness, I summarise the main points of procedure in Arbitration and then Litigation. It is worth noting that the processes are moving ever closer together, with some arbitrations being heard by retired high court judges.

11.3   Arbitration explained

Arbitration is a private alternative to litigation, which permits the parties to choose the adjudicator of their dispute. Arbitration must have a dispute that has been referred to an arbitrator who, once he has reviewed any evidence, will make a final and binding decision, subject always to review by the courts.

The process of reaching arbitration is a simple one. A conflict, dispute or difference arises, which cannot be settled by negotiation between the parties. Perhaps the parties attempt mediation, but this is usually not the case. One of the parties seeks a reference to an arbitrator and issues a notice of arbitration, seeking the appointment of an arbitrator. From then on, there are formalities to be observed.

11.3.1 Arbitration agreement

Because arbitration is a private system of dispute resolution, it has to be agreed between the parties. There are statutory arbitrations, but not in our industries generally. The parties can agree to arbitration at any time but, if there is no agreement, then one party can refuse to arbitrate and seek a court-based remedy.

Unusually, however, in the construction and engineering folds, contract conditions include arbitration agreements. Some are quite vague, some very specific, but all must be in writing. The JCT arbitration rules, for example, are very specific, and they give a detailed procedure and timetable for events. Once an arbitration agreement is agreed, either in the contract or later, then the parties must use arbitration. The clearer the agreement, the less likely the parties are to conflict over the agreement itself.

Unfortunately, some agreements are challenged as to their validity. Are they properly incorporated into the contract? Are they explicit? Do they actually mention the word arbitration? A good many conflicts, over the years, have arisen on this topic alone, notwithstanding the issues. The next stage of conflict occurs in the selection of an arbitrator.

11.3.2 Selection of arbitrator

As mentioned earlier, the parties may select the arbitrator. However, as the parties are already in dispute, they often dispute everything, even the name of the arbitrator. In an impasse such as this, the domestic contract forms in existence today permit selection by an outside body, such as the ICE, RICS or the Chartered Institute of Arbitrators.

On rare occasions, where the parties fail to agree on an arbitrator, and an outside body does not have the authority, the courts will appoint the arbitrator. The key to avoiding conflict here is to select a shortlist of arbitrators, which both sides would be happy with very early in the project, whilst disputes are just a possibility. Many disputes have become further embittered by arguments over the appointment of an arbitrator.

11.3.3 The Arbitration

The Arbitration will be governed by the statutes passed by the legislature, namely the current Arbitration Acts, with other incidental legislation having an influence.

The Arbitrator should be truly neutral, without an interest in the issue or in either party. Naturally, if the Arbitrator exceeds his powers or refuses to do his duty, he will be in danger of removal. Should the Arbitrator act fraudulently, then his powers to act can be revoked. Less obvious types of misconduct also provide grounds for dismissal, the most likely being 'dilatory performance'. Occasionally, one or both parties are frustrated by the failure of the arbitrator to deal with matters reasonably promptly. In these cases, it is up to one party or the other to act, to deal with the matter.

The Arbitrator has wide powers and, so long as he acts fairly, can exercise these powers without interference from the courts. Many important legal cases have arisen from challenges to an arbitrator's authority to exercise his wide powers. In most cases, the law courts have upheld the rights of arbitrators, for example, to 'open up and review certificates of the Architect', and to make 'supplemental awards' where losses could not be specifically supported one from another.

The Arbitrator, in the same way as a judge, has intermediate hearings for the giving of directions, and can order discovery, security for costs and other pertinent procedures to be addressed.

The main causes of conflict within arbitration again revolve around uncertainty. Where should it be held? Under what legal jurisdiction? Under which arbitration rules? Some plaintiffs will be unhappy with a timetable that continually slips because of a slow defendant and the acquiescence of the Arbitrator.

Conflict within arbitration is just as likely as it is within the project itself. This is because co-operation is necessary for a successful arbitration. If both parties can concentrate on the arbitral resolution of the conflicting issues, then perhaps arbitrations will be less tortuous.

The procedure in an arbitration hearing is a little less formal, in theory, than in the courts, but follows the same pattern, and this is covered later.

11.3.4 The Arbitrator's award

Procedurally, the award must be made by the Arbitrator himself, signed by him and witnessed by another. His award will be delivered to the parties and, in layman's terms, could be considered the equivalent of the judgement of a court.

From a practical point of view, the award should have the following facets:

- It should be clearly written and carefully drafted.
- It should be checked by the Arbitrator.
- It must decide, with finality, every matter referred.
- It must be certain as to its effect.
- It must be consistent throughout.
- It must be well reasoned.
- It must reflect the current law.

Occasionally, the Arbitrator is asked to make an interim award on a limited number of issues, leaving others until later. Once an interim award is made, it is as final as the overall award, and cannot be reopened. Once the final award has been made, the Arbitrator drops out of the picture. Any issues that the parties wish to argue or appeal further are now outside the Arbitrator's remit.

The award itself will include costs and interest, which should be awarded as they would be in the High Court. This is discussed in more detail later.

11.3.5   Appeal

There is a limited right of appeal to an arbitrator's award. Under the Arbitration Act, an appeal shall lie before the High Court on any question of law arising out of an award made on an arbitration agreement. If the appeal were to succeed, then the court may vary or set aside the award, or remit the award to the Arbitrator, along with the court's opinion on the question of law raised. There are far fewer appeals under the more recent Arbitration Act than previously, and this makes conflict arising out of the award itself a less worrisome problem.

Whilst this is, by no means, a detailed review of arbitration, it has sought to show where conflicts can arise, even after a matter is referred to arbitration. The alternative, if there is no arbitration agreement, is to refer the matter to the courts for resolution.

11.3.6   Arbitration panels

Whilst domestic disputes will often be heard by a single arbitrator, most of my arbitrations are international and are governed by either the ICC or UNCITRAL rules. Most will have three arbitrators, with each party choosing one arbitrator and the third being chosen by the two arbitrators or the ruling body.

The procedures will otherwise be largely the same, with the Chairman being the main contact for all correspondence.

11.4   Litigation explained

As we have noted above, conflicts in the construction industry do become more entrenched as time goes by, and if the conflict is inevitable, a path must be chosen for its resolution. Very often the path chosen is litigation, the process that leads from the issue of a writ to a hearing and then on to a judgement. If a writ is raised when an arbitration agreement is in place, the courts will usually grant a stay of prosecution and refer the parties back to the arbitral process incorporated within their contract. However, often no such arbitration agreement is incorporated into the contract, and the matter proceeds to litigation.

One of the failures of the legal process has been the speed with which solicitors have been prepared to issue writs. In the past, writs have been issued as a simple debt collection tool, the solicitor and his client hoping that the recipient would concede in order to avoid the legal process. This is a form of coercion, or perhaps even blackmail, and generally people react badly to threats of this type. Thus, a clumsy attempt at debt collection brings in a full defence and a counter-claim, opening the conflict up further. After this point has been reached, the plaintiff cannot simply withdraw his writ and forget the matter. If he did so, the counter-claim would be found against him along with the costs. The plaintiff must fight on, and another case spins inexorably towards a hearing. This is unnecessary, but not uncommon.

In recent years, we have seen ADR being used by solicitors more readily, in an attempt to avoid unnecessary litigation. There are, however, still many writs being issued unnecessarily, often by smaller practices with limited experience of complex construction disputes. If a dispute does require litigation, then we need to know how the procedures will operate, and so the remainder of this chapter is an overview of these procedures.

11.4.1   The Writ

Generally speaking, the disputes or conflicts with which we are concerned are better dealt with by specialist judges. We do have such a bench of specialist judges who operate within the Technology and Construction Court (TCC). With their own building, specially designed and built for these complex civil cases, the judges deal largely with construction, engineering, technology and shipbuilding cases. This précis, therefore, concentrates on the procedures for the TCC.

Initially, the originating summons or writ should be endorsed with the letters 'TCC' and the case will be listed specifically for that court. One of the advantages of dealing with the case in the TCC, over and above the High Court, is that the Judge deals with the whole case, not just the hearing. In the High Court, much of the interlocutory work (this is explained later) is done not by the Hearing Judge but by a High Court Master. An appeal from a master's decision goes to the Judge and then to the Court of Appeal. An appeal from an interlocutory decision in the TCC goes straight to the Court of Appeal.

If the writ begins life in the High Court, it can always be transferred at a later date, and indeed the High Court may decide this. Even within the OR's court, one OR can transfer a case to another, giving flexibility and so reducing waiting times. Once the writ is issued, matters proceed apace.

11.4.2   Pre-Hearing

Before a full hearing of the case, probably months away, a host of formalities have to be dealt with, and the Judge deals with these matters in interlocutory hearings. Following the Pre Trial Protocol, the Judge will probably insist that the parties try mediation before proceeding to a full hearing. Such mediations do have a reasonable record of success.

The pleadings are the real starting point of the procedure. The plaintiff states his case concisely and clearly, citing the facts upon which he relies. The evidence is not provided at this time, nor should the legal matters be pleaded. A defendant receiving the pleading should know from the pleadings the case he has to answer, and then set about doing so in his defence and counter-claim. The plaintiff may then reply to the defence and defend against the counter-claim.

Sometimes, at this stage of the proceedings, it will be clear that there is no real defence and that the defendant is simply refusing to pay what seems clearly due. If this is the case, the plaintiff will seek, under the Rules of the Supreme Court (RC), an Order 14 for Summary Judgement. This is a rapid method, in a legal timescale, of recovering a debt undoubtedly due. If the defendant cannot convince the court that he has an issue that ought to be tried, then immediate judgement can be given in favour of the defendant. The judgement can be for the whole or part of the money claimed. Order 14 has often been used to secure payment of certified monies, which have not been paid. If successful at this stage, then the plaintiff often secures his money and the matter proceeds no further, but if it does, there is a long way to go.

Construction and engineering disputes are complex, and it is comparatively rare to find a case fully pleaded in the first instances. To aid parties to fully understand the case against them, each party is entitled to seek further clarification by way of Further and Better Particulars and Interlocutories. Thus, by asking for more details of a particular claimed item, or by specific questioning, both sides can build a sound case and avoid raising items that are not in dispute. Orders to this effect are given in hearings by the Judge.

From time to time, the parties go to the court and seek directions and a timetable for the litigation. The Judge gives these directions and issues a reasonably flexible timetable. The parties are given a reasonable time in which to provide revised pleadings, further and better particulars, and so forth. If, for one reason or another, the Judge believes that one party is not reacting with appropriate speed, he can issue an 'unless order'. Then, unless the documents are provided by the date cited, the defaulting party may lose their right to claim.

Some of the more popular orders have been discussed earlier, and these attempt to narrow the triable issues, thus saving time, effort and, more importantly, cost. They are:

- *Scott Schedules*: Facts and figures on a comparative spreadsheet – but not legal arguments:
- *Expert Witnesses*: To provide unbiased, expert opinion as to quantum, planning, liability, etc.

When all of the issues have been crystallised, the parties are ready for a hearing.

11.4.3 Hearings

The procedures for hearings are covered below, but it is worth noting that there are two types of hearing. A preliminary hearing may be held to deal with a major point or issue upon which the whole case may turn. For example, Client A insists that Contractor B is in breach of contract, and lists his losses.

Contractor B is insistent that no contract exists, therefore, no contract, no breach, no case. Clearly it would be in the best interests of all parties to have the contract/no contract issue resolved first, as the resolution of that issue will determine whether or not the case needs to continue.

By the process of elimination, the reader will have gathered that a full hearing will deal with all matters left unresolved by preliminary hearings.

Before the trial, the solicitors will have created a core bundle of relevant documents and this bundle will contain all documents upon which plaintiff and defendant rely. Also, the witnesses will have given statements to their respective solicitors relating the evidence they will give in court. These witness statements are usually given to the Judge long before the hearing, and he will not expect to hear new evidence being given by a witness from the witness box that was excluded from their statement. By the hearing date, the expert witnesses will have reported and discussed the issues. They may even have provided a joint report, which will save valuable court time.

In the weeks before trial, everyone spends a great deal of time preparing, especially the legal representatives, and their time is expensive. To cover these costs, the solicitors will usually be on an hourly rate, and the barristers will seek a retainer to cover his preparation costs, and Daily Refresher to cover their daily fees and expenses. The week before trial is a very expensive week indeed for both parties.

In the hearing itself, the barristers give their opening statements, explaining to the judge, and often to themselves, exactly what the case is about. The plaintiff goes first, and then the defence.

As witnesses are called, they are led through their statements kindly and supportively by their own side's counsel. They are then taken through their evidence again, less kindly, by counsel for the opposition. Their counsel then may re-examine, to clarify any confusion caused by the opposition. These three steps are:

1.   Examination-in-chief
2.   Cross-examination
3.   Re-examination.

By this approach it is hoped that the truth will be established, and usually it is. Judges are very astute, and can spot dishonesty or evasion very quickly. After all, they have been doing this on most days for many years.

When all of the evidence has been heard, both evidence of fact and of opinion, Counsel from both sides give their closing remarks. The Judge retires and writes his judgement. This is not always a rapid process and, in some cases, the judgement has appeared between three months and a year later, although three months is probably the better guide.

The judgement summarises the dispute, highlights the issues and gives the reasons for the Judge's decisions. The damages, if any, are awarded, and costs are awarded and taxed. The damages are the value that the Judge places on the loss the plaintiff has suffered and, under High Court Rules, these are added to the damage interest and costs. Interest is levied at the rate laid down, for the time being, by the court, and usually runs from the time the plaintiff was deprived of his money to the date of judgement.

Costs accrue on both sides, and these are taxed by a taxing master. This simply means that the courts look at the costs incurred, allowing those that are reasonably incurred and disallowing those that are not. Rarely, if ever, do the parties receive their full costs; often as little as 60% is recoverable. The Judge decides upon the balance of rights and wrongs, how the costs should be split, and generally, unless there are exceptional reasons, he will award the full costs of his case (after taxing) to the successful plaintiff or defendant. In some cases, losing the battle on costs is more ruinous than losing the sum originally in dispute, and this should be recognised by any prospective litigant.

So, our conflict has reached the end of the line. Or has it?

So far as a decision goes, it has, but it is a rare occurrence for both parties to shake hands after a long trial and mean it. The feelings, disappointments and perceived injustices will go on hurting for a long time. Some are never forgotten. Our aim must always be to avoid formal resolution methods wherever possible.

Where do we go from here? Can we save construction from itself? If we can, then it will be as a result of following the advice already given in this book on avoiding, reducing and managing conflict. There are sunnier skies ahead, however, and in our concluding chapter, we look forward to anticipate what the future holds for conflict in the industry.

12 Conflict in Changing and Challenging Markets

In the previous chapters, we have sought to set out the symptoms, causes and effects of conflict. We then addressed the behaviours and techniques, which need to be developed to avoid or reduce conflict. The next stage was to examine how inevitable or ongoing conflicts could be managed. Finally, we reviewed the methods of containing and resolving conflict.

Whilst all of this has been necessary and worthwhile, perhaps the greatest emphasis should be placed on the prevention of conflict, and examination of how we can prevent conflict from arising in the future. The details are elsewhere in the book, and so this chapter is dedicated to broad concepts and ideas, which can be addressed to alleviate the pressure on the industry to conflict.

Purely for convenience, we have divided this chapter into headings. However, it is not intended to suggest that the earlier headings are more important than the later, or vice versa. The harmonisation of construction, and its related industries, will come as a result of examining all possibilities and finding useful contributions from each of them.

The question for this final chapter is essentially a simple one:

Is there real hope in the future for a general reduction in Construction Conflict?

12.1 Will people change?

You will have noticed that the most intractable conflicts revolve around people. Is there something that we can do to affect this problem? Perhaps the answer is to reduce the number of people in the industry, who have conflictive or adversarial tendencies. How could this be done? Is this already happening naturally as society changes?

Conflicts in Construction: Avoiding, Managing, Resolving, Second Edition. Jeffery Whitfield.
© 2012 John Wiley & Sons, Ltd. Published 2012 by John Wiley & Sons, Ltd.

12.1.1 Social change

If you had read this chapter in the first edition of this book in 1994, you would have seen a forecast that the industry would become less conflictive. This was not a prophecy that demanded an unusual degree of prescience; it was self-evident that we could not go on as we were. Within two years of publication, we saw the rise of partnering, the intervention of adjudication on interim payment disputes and new legislation intended to reduce conflict. We also saw construction students leaving university having studied conflict, its causes and management. This book was widely used as a text book for that very purpose.

The industry has undoubtedly benefited from the new blood, the new ideas and the new approach being proselytised by the upcoming generation. Society in general has also softened markedly; people are less complaining, their immediate needs having been more amply satisfied than in previous generations. We now recognise that our personal, political and corporate behaviour has an impact on other cultures, and we seem to care more about minimising our negative impact on the world.

With a more considerate global society, perhaps we will see a more caring national culture too. The 1960s and early 1970s were said to have preached caring without being able to achieve it. The 1980s in the UK have been described as the selfish years, and in the 1990s and 2000s, we moved towards a more thoughtful approach to others and their needs. If this can be true nationally and globally, it can be true locally and personally. One truly charitable and caring person will always affect many lives.

We probably should have expected to see these changes as we have been taught for many years that, as people begin to satisfy the basic needs of human existence, such as food, shelter, health and well-being, their minds and bodies are freed to move higher up Maslow's hierarchy and seek after more comprehensive fulfilment. Almost inevitably, the greatest fulfilment comes from working with people in peaceful surroundings.

As our population moves away from seeing life as a game of survival, we should also see less aggressive dissent. If this happens, we will begin to see more co-operation and less conflict. At least this is the hope.

Clearly, general changes in society towards a less stressful lifestyle would help to harmonise society generally and construction particularly.

12.1.2 Education

We discussed earlier the way in which education, formal and informal, affects our attitudes towards conflict. Our colleges and universities have recognised conflict as a problem in the industry.

The government continues to sponsor research and introduce practical measures, which aim to reduce conflict in construction. The universities are teaching a new generation of construction graduates a better way of working. These graduates will, in the years to come, be leaders and managers in the industry. They will, hopefully, still believe in, and practice, the less contentious approaches taught in their youth. Those of us who have been in the industry for some time may be tainted by our own experiences, and we must be careful not to pass our negative behaviour on to this new generation. That is our responsibility.

12.1.3 Women in construction

There is little doubt that women are less aggressive than men, generally. So, perhaps with more women in the industry, we will see an end to the excessive 'macho' behaviour encountered presently. As women introduce a more people-centred emphasis into the industry, we will all benefit from the results. People have always been more important than things, and women recognise that to think like this shows not weakness but strength. Compassion, of course, has its limits in trying to get a job done and a project completed, but a balance can be achieved. Men in the industry must recognise in women the qualities that are positive contributors to conflict reduction, and avoid suggesting to women that to be successful in construction they need to be pseudo-men.

Let us instead use the respective strengths of men and women to improve the industry, rather than trying to mould all new entrants into an outmoded model, which never worked terribly well even when new.

12.1.4 The New Man

We have heard much talk in recent years about the 'New Man'. He is described as a more caring, more considerate and empathetic individual. The New Man is in touch with his emotions, and is aware of his purpose and role. He also feels okay about himself, which enables him to feel okay about others. Because our New Man understands himself, he is able to do what are viewed as less masculine tasks – such as looking after the children, housekeeping and caring for the sick – without feeling emasculated.

The New Man knows that masculinity is derived from who you are, not from how you behave. If the perception is true, and there are some positive signs to suggest that it has a basis in truth, then perhaps we can look forward to men proving their macho capabilities in their chosen sport and not by conflicting important construction projects.

12.2 Will contracts change?

If we can control and reduce the conflictive tendencies of individuals, then obviously we have addressed a major problem. But what is it that gives people a reason to conflict?

Over the last 20 years in construction and engineering, worldwide and not just the UK, contracts have moved in the direction of being fairer with risks being better balanced. It has taken a great deal of energy from a great number of individuals to achieve this. Perhaps the next changes will be intended to make those contracts less adversarial and more investigative.

Think of your favourite contract form, and consider the clauses that give extensions of time and more money to the contract. What do these clauses require? Is it:

- a notice from the Contractor;
- an acceptance or rejection by the client;
- arbitration, if you cannot agree.

Ask yourself why it is not rather:

- a notice that a problem has arisen;
- investigation by both parties as to causes;
- agreement as to the impact of the problem;
- agreement as to how to overcome/mitigate the problem;
- immediate third-party help, if agreement is not reached.

In practice, I have seen the second procedure followed on site by individuals, who know that co-operation is the way to complete on time and on budget, but it is not how the contract reads.

Lawyers and construction professionals need to be innovative in the field of preparing contract documentation, to ensure that conflict is avoided wherever possible. Here are some *improvements*, which a class of young site quantity surveyors came up with in a training session:

- Use plain English and short sentences.
- Avoid legal jargon.
- Make it clear what each party has agreed to do.
- Keep the important clauses together (those to do with the project's success) and away from sundry issues, such as VAT and the Finance Act.
- Make the contract a maximum of two sides of A4.
- Have it written by trainee quantity surveyors.

Perhaps contracts cannot be simplified sufficiently to avoid misunderstanding altogether, but they can certainly be improved in terms of co-operation.

12.2.1 *Partnering*

There is little doubt that Partnering creates an atmosphere where greater co-operation can flourish on construction projects, although some contractors believe that this harmony is obtained at the cost of the Contractor constantly giving in. Most large clients, such as power generators, supermarkets, hospital groups and developers, have established links with contractors and sub-contractors on whom they can rely. The ongoing relationship created by such partnering has two significant benefits when it comes to reducing conflict, namely:

- an interest in an ongoing relationship, so that small disputes are put into their proper perspective;
- the relationship is based upon trust and co-operation.

Partnering will not exclude the possibility of conflict, but it will certainly encourage the parties to minimise the impact that the conflict has upon the project and upon the future relationship.

12.3   Does Europe have anything better to offer?

The very different approach to contracting in Europe is both culturally and legally motivated. In Europe, statutory law covers a greater proportion of the commercial marketplace than in the UK. Our domestic laws enable us to enter into quite onerous contracts, and will not help us when we balk at our obligations. In Western Europe, a plethora of laws protect the smaller companies and individuals from contracted exploitation.

Culturally, our European counterparts have always preferred partnering and joint ventures to the traditional hierarchy of UK construction projects. It would seem that this should reduce conflict dramatically, but unfortunately this is not so. They simply conflict over different things, for example, who suffers the losses, or who takes the profits.

It is in the area of conflict resolution that the Europeans have an advantage over both the UK and the US construction industries.

Firstly, they are reluctant to turn to the law for a remedy, and so will almost always seek after a negotiated settlement. Compromise is a way of life to many mainland Europeans, and this attitude permeates their construction processes too. After a particularly difficult negotiation in France recently, I had to explain to a French sub-contractor that he was not going to be paid the sum he claimed. The director at the negotiations gave a typical Gallic shrug of the shoulders and said:

C'est la vie, there will be another job tomorrow.

Secondly, their legal system is not adversarial but is investigative, which helps the parties to remain on friendly terms. After a quite hotly-contended hearing, there can be a lot of shouting in French courts. In one instance, the plaintiff who lost his case walked towards our client, and I thought briefly that violence was on his mind. The two opponents then hugged one another, and agreed to go to a café together for a beer. Perhaps there is something that we can learn from Europe, after all.

12.4   Conflict in changing market conditions

After a period of unparalleled investment in our infrastructure in the last decade, we now face a reduction in public spending. In addition, we are faced with potential reductions in bank lending and private investment. The housing market, particularly in the apartment sector, has suffered badly. New starts are down and unsold dwellings by the thousand attract little or no serious interest.

In any public spending moratorium, capital spending goes first, and we can expect some difficult times. Work will be available overseas for companies and for individuals, but the rewards that were achievable in the past are now not achievable.

It is possible that in due course conflicts will increase again as cash becomes scarce, developers will look for ways to reduce construction costs and contractors will chase turnover, and profits will be made by exploiting ambiguity, change and error. In the UK and Middle East, there already exist pent-up conflicts, just waiting for money to flow into the funders' pockets so that they become worth pursuing.

Luckily there is another way. Clients and developers can be open and honest about their future programme of work and their budgets. Contractors, clients and their professionals can work together to try to obtain the maximum value for money from the projects that do proceed. Then, when the governments see that money is being spent wisely again, perhaps they will grow the economy and invest in public service capital projects and national infrastructure.

We will face difficult times, but an increase in conflict will never make them easier.

12.5   More conflict or less conflict – you decide!

We have looked, with some optimism, at the future and how current trends can assist in reducing conflict. However, it would be remiss of me not to mention that, just as no-one else can make us angry (becoming angry is our decision), then no-one else can force us to conflict, either. Remain unemotional, clear headed and remember, when under attack, God never made bad people – he made good people who behave badly.

If we can change the way we think, we will be changing the way we act. This will, in turn, change the way that others react, and a step forward will have been taken.

Of the people I know in the construction industry, many are ingenious at finding new and refreshing ways of conflicting. I hope that this book will help us to channel that ingenuity into finding exciting ways of avoiding and dealing with conflict instead.

Index

Conflicts in Construction: Avoiding, Managing, Resolving, Second Edition. Jeffery Whitfield.
© 2012 John Wiley & Sons, Ltd. Published 2012 by John Wiley & Sons, Ltd.

www.ingramcontent.com/pod-product-compliance
Lightning Source LLC
Chambersburg PA
CBHW080941260125

20788CB00015BA/178